D1577990

Exploring
Algebra 2

with

PAUL KUNKEL

STEVEN CHANAN

SCOTT STEKETEE

Key Curriculum Press
Innovators in Mathematics Education

Project Editor:	Scott Steketee
Editorial Assistants:	Aaron Madrigal, Nina Mamikunian
Contributing Writers:	Lyubomir Detchkov, Ralph Pantozzi, Kevin Thompson, Ned Diamond, Dan Dudley, Nathalie Sinclair, Dan Bennett, Eric Bergofsky
Reviewers:	Daniel Scher, Molly Jones, Dan Lufkin, Marsha Sanders-Leigh, John Threlkeld, Pat Brewster
Math and Sketch Checker:	Bill Zahner
Correlations:	Nat Lewis
Editorial Production Manager:	Christine Osborne
Production Editor:	Angela Chen
Copyeditor:	Tom Briggs
Production Coordinator:	Jennifer Young
Compositor:	Graphic World
Art Editor:	Jason Luz
Technical Art:	Integra
Text Designer:	Marilyn Perry
Cover Designer:	Jensen Barnes
Cover Photo Credit:	© David Stoecklein, Corbis
Printer:	Versa Press
Managing Editor, Media Programs:	Joan Lewis
Vice President Editorial and Production:	Casey FitzSimons
Publisher:	Steven Rasmussen

®The Geometer's Sketchpad, ®Dynamic Geometry, and ®Key Curriculum Press are registered trademarks of Key Curriculum Press. ™Sketchpad is a trademark of Key Curriculum Press. All other brand names and product names are trademarks or registered trademarks of their respective holders.

Limited Reproduction Permission

© 2007 Key Curriculum Press. All rights reserved. Key Curriculum Press grants the teacher who purchases *Exploring Algebra 2 with The Geometer's Sketchpad* the right to reproduce activities and example sketches for use with his or her own students. Unauthorized copying of *Exploring Algebra 2 with The Geometer's Sketchpad* or of the *Exploring Algebra 2* sketches is a violation of federal law.

Exploring Algebra 2 **Sketches CD-ROM**

Key Curriculum Press guarantees that the *Exploring Algebra 2* CD-ROM that accompanies this book is free of defects in materials and workmanship. A defective CD-ROM will be replaced free of charge if returned within 90 days of the purchase date. After 90 days, there is a $10.00 replacement fee.

Key Curriculum Press
1150 65th Street
Emeryville, CA 94608
510-595-7000
editorial@keypress.com
www.keypress.com

10 9 8 7 6 5 4 3   10 09 08   ISBN: 978-1-55953-799-5

Contents

5: Algebraic Transformations

6: Other Functions

7: Trigonometric Functions

8: Probability and Data

9: Vectors and Matrices

Exploring Algebra 2 with The Geometer's Sketchpad
© 2007 Key Curriculum Press

CD-ROM Contents

Be sure to check the **Read Me.gsp** file on the CD-ROM for activities, sketches, and tools that may have been added after the book went to press.

Activity Sketches

The sketches that support the activities are included in this folder. For instance, the sketch for the Introducing Dynagraphs activity is located on the CD-ROM in this location: **Activity Sketches | 1 Functions | Introducing Dynagraphs.gsp.** For most of the activities, there are also presentation sketches that you can use to present the activity in a whole-class setting.

Activity PDFs

The PDF documents in this folder include every activity in this book. These documents will make it easy for you to print out activity sheets for your students to use. (To read PDF documents, use Adobe® Reader®, a free download from www.adobe.com.)

Chart of Activities.pdf

This PDF document contains the Chart of Activities in electronic form, making it easy for you to choose activities for your class. For every activity, this chart lists the student audience, Sketchpad level, estimated time, setting, and description.

Correlation PDFs

The activities in this book have been correlated with several Algebra 1 and Algebra 2 textbooks. You will find the correlation tables (in the form of PDF documents) in the **Correlation PDFs** folder on the CD-ROM. Check this folder to determine whether a correlation for your textbook is included.

Projects.pdf

Sketchpad projects can play a special role in engaging students and developing their mathematical thinking and competency. This PDF document contains a list of projects that you can suggest or assign to students.

Supplemental Activities

This folder on the CD-ROM contains activities that are not printed in the book. As of this writing, the following activities are included; check the CD itself to see if others have been added.

Quadratic Intercepts

Students learn the sum-of-roots and product-of-roots formulas for quadratic functions and use them to derive a quadratic function from three intercepts.

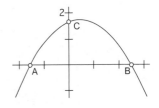

Function Transformation Game

Students match the graph of a mystery function by choosing a parent function and applying transformations to it.

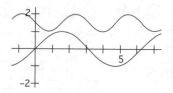

Point Field

Students use a prepared point field to investigate linear combinations in the form $ax + by$.

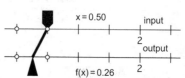

Build Your Own Dynagraph

Students or teachers create their own dynagraphs similar to the dynagraphs used in the activities in Chapter 1.

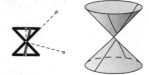

Supplemental Sketches

This folder contains sketches that are useful for many purposes: viewing conic sections in three dimensions, graphing trigonometric functions, and so forth.

Conic Sections.gsp

Students can adjust the angle at which a plane intersects a cone, and look at the intersection from any desired three-dimensional viewpoint. Have students experiment with this sketch, or use it in a short presentation, to communicate the original meaning of the term *conic section*.

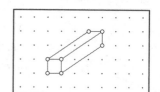

Geoboards.gsp

Students or teachers can create a variety of geoboards and grids, including rectangular geoboards, isometric geoboards, and rectangular grids displaying optional axes and numbers. They can then construct points that snap to the defined grid.

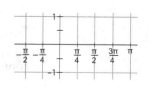

Trig Coords.gsp

Students can use the coordinate system in this sketch to plot functions using radians, expressed as fractions of π, on the x-axis.

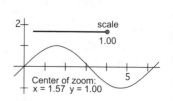

Zooming Coordinate System.gsp

Students can use this coordinate system to plot functions and then zoom in or out on the resulting graphs.

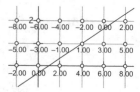

Exploring Algebra 2 with The Geometer's Sketchpad
© 2007 Key Curriculum Press

Supplemental Tools

This folder contains tools that are useful for many purposes by either teachers or students.

Axis Tools.gsp

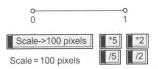

The Axis Scale tool creates an easily adjustable scale measurement that you can use to create coordinate systems, dynagraphs, and so forth. The Dynagraph tool uses a scale measurement to create a nicely formatted dynagraph.

Pi Fraction Reducer.gsp

$$n = 9 \qquad \frac{3\pi}{4}$$
$$d = 12$$

The first tool reduces fractions of π starting with a numerator and denominator. For instance, if the given values are 9 and 12, $9\pi / 12$ will be reduced to $3\pi / 4$. The second tool rounds an angle measurement to the nearest fraction for a given denominator. For instance, if the given values are an angle of 0.72π and a denominator of 4, the result will be $3\pi / 4$.

Inequality Tools.gsp

Use these tools to create graphs of inequalities, such as $y < 3x - 4$ or $x \geq y^2/2$. These are the same tools used in the activity Graphing Systems of Inequalities.

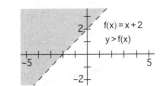

Coord Sys Scale Controls.gsp

Use these tools to create coordinate systems with scaling controls so that it's easy to adjust, for example, a scale of 20 pixels/unit to 50 pixels/unit. The first tool adds scaling controls to an existing coordinate system, and the second creates a new coordinate system with scaling controls.

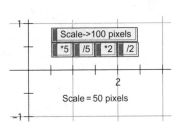

New Sliders.gsp and Advanced Sliders.gsp

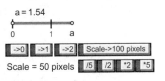

Use these tools to create a variety of sliders. Some have fixed scales that make it easy to drag the slider to values like 1.00 or 0.50, and others have adjustable scales that you can easily set to scales like 20 pixels/unit, 50 pixels/unit, or even to the same scale as an existing coordinate system. Also included are integer sliders and circular sliders for angle values.

Graphing Toolkit.gsp and Graphing Toolkit.pdf

The PDF document describes how to use the tools in the sketch to graph various functions and perform some of the same operations supported by graphing calculators.

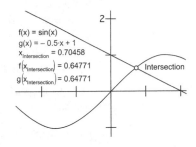

Using These Activities

Using the activities in this book, intermediate and advanced algebra students can directly create and manipulate algebraic objects and see the effects of their manipulation. Students understand, internalize, and retain algebra concepts through active engagement with visual models.

These activities allow students to

- drag a marker on an input axis to change the value of x and observe the resulting change in the value of the function on the output axis

- experiment with various forms of quadratic expressions, seeing how the graph changes as different parameters change

- translate, stretch, and shrink functions by manipulating sliders or parameters

- compare the unit circle definitions for the trigonometric functions with the triangle definitions

Through these and similar experiences, students encounter the excitement of algebra with unprecedented clarity and concreteness.

When to Use an Activity

These activities are designed to be independent activities, so that you only need to use the ones that fit your particular course of study. You don't have to present them in sequence, even though we've organized them into numbered groups. You should use them in whatever order best matches your own curriculum and textbook. Some activities may be easier for students if they've already done a related activity; such instances are described in the Activity Notes that accompany each activity.

You don't need to find extra time to fit these activities into your course of study because dynamic exploration actually reduces the amount of time that students need to master a particular topic.

The main audience for these activities is Algebra 2 students, and every activity is useful in one Algebra 2 course or another. Because learning is spiral and the boundaries blur between courses, many activities are also useful for Algebra 1 students, and others for Precalculus students. Similarly, you may find useful for your advanced algebra students a number of the activities in the companion volumes *Exploring Algebra 1 with The Geometer's Sketchpad* and *Exploring Precalculus with The Geometer's Sketchpad*.

The Chart of Activities on pages xiii–xvi gives a brief description of each activity, including the intended audience, the Sketchpad level, the time required, the setting (described below), and the objective. Refer to this chart to identify appropriate activities quickly.

The CD-ROM in the back of the book contains correlations of these activities with a number of popular textbooks. The correlation tables are a helpful resource when you want to pick out an activity that is appropriate for a particular lesson.

© 2007 Key Curriculum Press

Components of an Activity

The activities in this book contain several components.

- **Student Activity Sheets:** These blackline masters contain steps for students to follow and questions for them to answer. Duplicate them either by copying the printed page in this book or by printing the appropriate pages from the PDF file included on the CD-ROM. Every activity begins on a right-hand page, making it easier for you to copy the printed pages and to keep the pages organized after you tear them out.

- **Activity Sketches:** These sketches, designed for students to use during the activity, are included on the CD-ROM. In some activities, students manipulate the sketch by dragging objects and by pressing buttons; in others, students use the activity sketch as a jumping-off point for doing their own constructions. Activities that begin with a new empty sketch may not include an activity sketch.

- **Presentation Sketches:** These sketches, designed for presenting a particular activity to the entire class, are also included on the CD-ROM. Many activities include separate presentation sketches; for other activities, the activity sketch also serves as the presentation sketch.

- **Activity Notes:** Each activity includes notes describing the activity, providing suggestions for using it, and giving detailed answers to all the questions in the student activity sheets. These notes begin by specifying the mathematical objective of the activity, the student audience, any mathematical prerequisites, the level of difficulty, the expected time required, and the settings in which the activity is useful.

- **Presenter Notes:** Most activities include either a section in the activity notes describing how to do a whole-class presentation or a separate page of presenter notes containing detailed steps to follow and questions to ask during a presentation.

Instructional Settings

The activities in this book are flexible enough that they can be used in a variety of instructional settings:

- **Paired/Individual Activity:** In this setting, your entire class works in pairs or individually at a number of computers. This setting requires either a computer lab or a classroom with many computers available, and gives students the most direct experience with the algebra they are exploring. It's best to group two, or at most three, students at each computer to promote interaction, discussion, and cooperative problem-solving. (Some students may be "keyboard hogs." Halfway through the activity you should direct students to switch roles so that a different student handles the keyboard and mouse.)

To use an activity in this way, make copies of the student activity sheets and pass them out. It can be useful periodically to call the whole class together to discuss what they've done so far and to deal with any questions or difficulties they've encountered.

If you have a projection system available, you may want to demonstrate briefly certain techniques or steps, or to answer common student questions. Avoid the temptation of having students imitate your work on their computers, and don't spend too much time showing them what to do. Students benefit most from what they observe or figure out for themselves. When you demonstrate using a presentation system, consider having a student actually operate the computer.

Reserve time at the end of the activity for students to discuss and summarize as a whole class what they have done and observed.

- **Small-Group Activity:** In this setting, three to six students work together at a single computer. If you have a single-computer classroom, you can use this setting for enrichment or review for a single group of students. If you have more computers, you can have several groups working at once, either on the same activity or on different activities. Activities that work as Paired/Individual Activities generally also work well as Small-Group Activities. (A few Paired/Individual Activities that rely more heavily on direct student interaction with Sketchpad are not recommended for small groups. These are noted in the Chart of Activities.) The small-group setting also works well for independent projects; see **Projects.pdf** on the CD-ROM for suggestions.

- **Whole-Class Presentation:** In this setting, you use the Sketchpad-based activity and a single computer with a projection device to present a lesson and elicit student discussion. Many of the activities in this book have a detailed Presenter Notes page to guide you in your presentation, and most of the rest have a section at the end of the Activity Notes that describes how to use the activity as a presentation. (The Chart of Activities indicates which activities are recommended as Whole-Class Presentations and which have Presenter Notes.)

A whole-class presentation is not simply a Sketchpad-based demonstration that students watch. Student questions and discussion form an important part of the presentation, and you should look for ways to encourage participation. Consider having a student operate the computer during the presentation, allowing you to interact more effectively with the class and to encourage student questions and discussion. A rhythm and flow between class discussion and computer-based manipulation of an algebraic model can be wonderfully effective for promoting student involvement and understanding.

- **Short Demonstration:** In this setting, use just a portion of an activity to introduce or summarize a particular lesson or to illustrate a particular fine point. This setting requires a single computer and a presentation device in your classroom. For example, a lesson might flow smoothly from a Sketchpad-based introduction into teacher explanation, student discussion, questions and answers, and a Sketchpad-based summary. It's often best to have a student volunteer operate the computer for these brief demonstrations. When a student answers a question or makes a point during class discussion, consider asking the student to demonstrate using Sketchpad. Many of the activities have detailed Presenter Notes and/or presentation sketches that will help you in the process of showing a brief portion of the full activity.

The Chart of Activities on pages xiii–xvi shows which activities are recommended for use in each of these settings.

Even if you have a presentation system in your classroom and consistent access to a computer lab, we suggest you vary the ways you use the activities to benefit students with different learning styles.

No matter how you use a particular activity, involve students as actively as you can in the mathematics. Ask them to describe what they've done and what they've observed, and ask "what-if" questions to encourage them to speculate and make conjectures.

Activity Levels

The level of Sketchpad difficulty for each activity is rated as Easy, Intermediate, or Challenging.

- **Easy:** An easy activity involves manipulation of a prepared sketch, without requiring students to do any construction themselves. Students can do the activity with no prior Sketchpad experience.

- **Intermediate:** An intermediate activity involves some limited construction, described in detail in the activity steps and margin notes. The activity will be easier for students who have some prior Sketchpad experience, either in a previous course or by doing several of the activities ranked Easy.

- **Challenging:** A challenging activity involves substantial student construction, often starting with a new sketch. Although these activities include clear instructions, students will find them easier if they are already comfortable using Sketchpad. We recommend that students have some construction experience (either from previous courses or from doing several activities ranked Intermediate) before undertaking a challenging activity.

The Chart of Activities on pages xiii–xvi gives the Sketchpad level of each activity.

Activity Management

All of the activities will be easier for students if they cooperate, so encourage students to help each other as they construct and explore. In addition, they can often answer their own technical questions by using Sketchpad's Help system.

Your classroom will run more smoothly if you establish an unobtrusive yet obvious way for students to indicate that they have questions. In Key Curriculum Press workshops, we often provide red plastic cups that participants can display prominently when they have a question, so that we see at a glance where to direct our attention.

Every activity has questions for students, and none of the questions has a yes/no answer. Some questions prompt students to practice their skills, some test understanding, and some are intended to provoke thought and reflection. In many ways the questions are the most important part of the activity sheets, and you can often use them as a bridge to move from the Sketchpad activity to a class discussion and summary of the mathematics you are teaching.

Encourage students to answer questions thoughtfully and in complete sentences. Student activity sheets don't provide spaces for answers, so have your students write their answers on paper, type them into a word processor or email program, or use the **Text** tool to type them into Sketchpad. Some teachers have students write answers on a handout listing the questions, making it easy to confirm that each student has answered all questions.

Students learn best by doing and conjecturing, by answering questions and asking them. We hope that the activities in this book can provide your students with a novel opportunity to engage in these processes and to enliven their algebra learning as they vary variables, transform functions, and create trigonometric and other functions.

Acknowledgments and Writer Biographies

This book includes (in revised form) a number of activities that originally appeared in *Exploring Algebra with The Geometer's Sketchpad* (Chanan, Bergofsky, and Bennett, Key Curriculum Press, Emeryville, CA, 2001). The current volume adds a large number of new activities and includes explicit support for whole-class presentations.

We'd like to thank Dan Bennett, Eric Bergofsky, Daniel Scher, and Nathalie Sinclair, who created activities for the earlier book, and Lyubomir Detchkov, Ralph Pantozzi, Kevin Thompson, Ned Diamond, and Dan Dudley, who wrote new activities for this volume.

About the Writers

Paul Kunkel taught secondary mathematics in the state of Washington and is now a mathematics tutor in Hong Kong. He is the creator of Whistler Alley Mathematics, a website for math students and hobbyists. Paul is a co-author of *Exploring Precalculus with The Geometer's Sketchpad* and *Exploring Algebra 1 with The Geometer's Sketchpad*.

Steven Chanan is a graduate of Stanford University. He taught secondary mathematics for four years before he joined Key Curriculum Press as a technical writer and editor working on Sketchpad-related publications.

Scott Steketee is a graduate of Harvard College and holds master's degrees in Mathematics Education and in Computer Science. He taught secondary mathematics and computer science in Philadelphia for 18 years. Since 1992, he has worked on Sketchpad software development, curriculum development, and professional development.

© 2007 Key Curriculum Press

Chart of Activities

Use this chart to locate appropriate activities for various lessons in your curriculum.

Activity	Pre-algebra	Algebra 1	Algebra 2	Precalculus	Statistics	Sketchpad Level	Average Time	Paired/Individual	Small Groups	Whole-Class	Short Demo	Presenter Notes	Description
1: Functions													
Introducing Dynagraphs		○	○			E	35	○	○	○	○	○	Students explore dynagraphs to develop a feel for functional relationships.
From Dynagraphs to Cartesian Graphs		○	○			E	45	○	○				Students make connections between symbolic, Cartesian, and dynagraph representations of functions.
Domain and Range		○	○			E	35	○	○	○	○		Students explore the domain and range of functions, including those with restricted domain or range, using dynagraphs and Cartesian graphs.
Function Composition with Dynagraphs			○	○		E	40	○	○	○	○	○	Students use dynagraphs to model composite functions.
Odd and Even Functions			○	○		E	35	○	○	○	○		Students explore odd and even functions using dynagraphs and transformations.
Inverse Functions		○	○			E	30	○	○	○	○	○	Students use linked dynagraphs to investigate inverse functions.
Functions Again and Again		○	○			I	35	○	○			○	Students define an iterated coordinate transformation on a point, and observe and draw conclusions from the orbit.
2: Functions and Relations													
Relations and Functions		○	○			E	35	○	○	○	○	○	Students explore the definitions of relation and function, and develop a vertical line test for functions.
The Circumference Function	○	○	○			I	35	○	○				Students measure, graph, and analyze the function that connects a circle's diameter and circumference.
Radius and Arc Length		○	○			E/I	35	○	○	○	○	○	Students explore the relationship between the radius of a circle and the arc length of a semicircle.
Functions in a Triangle	○	○	○			I	30	○	○				Students measure constructions in a triangle and investigate the relations and their graphs.
Functional Geometry		○	○			I	30	○	○				Students explore relations defined by geometric measurements and create graphs, explaining how they decided on the independent variable.
3: Systems													
Solving Systems of Equations		○	○			I	35	○	○	○	○		Students use rate information from two companies to find out which is cheaper for various moves.
Graphing Inequalities in Two Variables		○	○			E	45	○	○				Students use a prepared sketch to graph various inequalities in x and y.

Legend: E = Easy; I = Intermediate; C = Challenging

Activity	Pre-algebra	Algebra 1	Algebra 2	Precalculus	Statistics	Sketchpad Level	Average Time	Paired/Individual	Small Groups	Whole-Class	Short Demo	Presenter Notes	Description
3: Systems continued													
Graphing Systems of Inequalities			○			I	40	○	○	○	○		Students use a prepared sketch to solve systems of two and three inequalities.
Linear Programming: Swans and Giraffes			○			E	45	○	○				Students explore a linear programming problem, by writing constraint equations, defining the feasible region, and maximizing a quantity.
4: Quadratic Functions													
Parabolas in Vertex Form		○	○			E	45	○	○	○	○	○	Students graph parabolas using the vertex form.
Exploring Parabolas in Vertex Form		○	○			E	30	○	○				Students graph parabolas using the vertex form (open-ended).
Parabolas in Factored Form		○	○			I	45	○	○	○	○		Students investigate the relationship between the factored form of a quadratic function and its graph.
Parabolas in Standard Form		○	○			I	40	○	○	○	○		Students use the standard form to identify the behavior of the graph when a, b, and c are changed.
Changing Quadratic Function Forms			○	○		C	50	○	○				Students change quadratic functions between standard, vertex, and factored forms.
The Discriminant			○	○		I	35	○	○	○	○	○	Students calculate and explore the discriminant of a quadratic function.
Parabolas: A Geometric Approach		○	○			I	40	○	○				Students construct a parabola geometrically.
Parabolas in Headlights and Satellite Dishes		○	○			C	45	○	○				Students construct and explore a two-dimensional model of a parabolic reflector.
Conic Reflections			○	○		C	45	○	○	○	○	○	Students explore reflective properties of ellipses and hyperbolas.
Modeling Projectile Motion		○	○			I	40	○	○	○	○		Students make a Sketchpad model of a basketball's flight, and make the ball go through a basket.
5: Algebraic Transformations													
Translating Coordinates		○	○			I	40	○	○	○	○	○	Students translate points in and make connections between the coordinates of a point and its translated image.
Rotating Coordinates		○	○			I	35	○	○			○	Students explore coordinate rotation of figures about the origin by multiples of 90°.
Reflecting in Geometry and Algebra		○	○			E/I	35	○	○	○	○	○	Students explore algebraic associations between the coordinates of a point and its reflected image.
Stretching and Shrinking Coordinates			○			I	30	○	○	○	○	○	Students investigate the behavior of polygons when the x- or y-values of the vertices are multiplied by various constants.

Legend: E = Easy; I = Intermediate; C = Challenging

© 2007 Key Curriculum Press

Activity	Pre-algebra	Algebra 1	Algebra 2	Precalculus	Statistics	Sketchpad Level	Average Time	Paired/Individual	Small Groups	Whole-Class	Short Demo	Presenter Notes	Description
5: Algebraic Transformations continued													
Transforming Coordinates		○	○			I	35	○	○			○	Students perform elementary transformations in the coordinate plane.
Translating Functions			○			I	25	○	○	○	○	○	Students translate function graphs vertically and horizontally by adding constants to x- and y-values.
Reflecting Functions		○	○			I	23	○	○	○	○	○	Students reflect function plots across the axes and explore connections between algebraic and geometric transformations.
Stretching and Shrinking Functions			○			I	25	○	○	○	○	○	Students stretch and shrink function graphs vertically and horizontally.
Transforming Odd and Even Functions			○			C	30	○	○	○	○		Students explore the symmetry in odd and even functions.
6: Other Functions													
Absolute Value Functions		○	○			E	25	○	○	○	○	○	Students graph and explore the absolute value function, reviewing the point-slope form of linear functions.
Exponential Functions			○	○		I	35	○	○	○	○	○	Students graph exponential functions, examine their properties, and use them to model real-world applications.
Logarithmic Functions			○			I/C	45	○	○	○	○	○	Students explore the relationships between exponential and logarithmic functions.
Square Root Functions			○			I	35	○	○	○	○	○	Students explore the square root function and think about the conditions under which inverse relations are also inverse functions.
Rational Functions			○			I	35	○	○	○	○	○	Students explore rational functions as transformations of $y = 1/x$.
Modeling Linear Motion: An Ant's Progress			○			E	25	○	○	○	○	○	Students model linear motion using parametric equations.
7: Trigonometric Functions													
Right Triangle Functions			○			I	25			○	○	○	Students calculate ratios for right triangles, plotting the values to reveal the graphs of the trigonometric functions.
Radian Measure			○	○		I	25	○	○	○	○	○	Students explore the relationship between the length, radius, and central angle of an arc.
Unit Circle Functions			○	○		I	40	○	○	○	○	○	Students use a unit circle to define the trigonometric functions.
Unit Circle and Right Triangle Functions			○	○		I	25	○	○	○	○	○	Students compare the unit circle definitions and right triangle definitions of trigonometric functions.
Trigonometric Identities			○	○		E	35	○	○	○	○	○	Students use geometric relationships to justify trigonometric identities.

Legend: E = Easy; I = Intermediate; C = Challenging

7: Trigonometric Functions continued

Activity	Pre-algebra	Algebra 1	Algebra 2	Precalculus	Statistics	Sketchpad Level	Average Time	Paired/Individual	Small Groups	Whole-Class	Short Demo	Presenter Notes	Description
Law of Sines			○	○		I	20	○	○	○	○		Students explore the Law of Sines and develop a proof.
Law of Cosines			○	○		I	35	○	○	○	○	○	Students develop the Law of Cosines by exploring how the Pythagorean theorem fails for triangles without a right angle.

8: Probability and Data

Activity	Pre-algebra	Algebra 1	Algebra 2	Precalculus	Statistics	Sketchpad Level	Average Time	Paired/Individual	Small Groups	Whole-Class	Short Demo	Presenter Notes	Description
Normal Distribution			○	○	○	E	35	○	○				Students use a random distribution to explore the normal density curve.
Permutation and Combination			○		○	E	45	○	○				Students explore permutations and combinations of a given set of objects.
Box and Whiskers		○	○	○	○	E	45	○	○			○	Students change data and explore the effects on a box-and-whiskers plot.
Fitting Functions to Data			○	○	○	I	40	○	○	○	○	○	Students transform functions to fit data and use a least-squares calculation to judge how good the fit is.

9: Vectors and Matrices

Activity	Pre-algebra	Algebra 1	Algebra 2	Precalculus	Statistics	Sketchpad Level	Average Time	Paired/Individual	Small Groups	Whole-Class	Short Demo	Presenter Notes	Description
Introduction to Vectors: Walking Rex		○	○			E	25	○	○	○	○	○	Students explore vectors, learning the connection between two ways to describe vectors.
Vector Addition and Subtraction		○	○			E	25	○	○	○	○	○	Students add and subtract vectors, and explore the commutativity of these operations.
Solving Systems Using Matrices			○	○		I	25	○	○	○	○	○	Students solve a system of equations expressed as a single matrix equation.

Supplemental Activities (on CD-ROM)

Activity	Pre-algebra	Algebra 1	Algebra 2	Precalculus	Statistics	Sketchpad Level	Average Time	Paired/Individual	Small Groups	Whole-Class	Short Demo	Presenter Notes	Description
Quadratic Intercepts		○	○			I	30	○	○	○	○	○	Students derive a quadratic function from the y-intercept and the two x-intercepts.
Function Transformation Game			○	○		E			○	○			Students match the graph of a mystery function by choosing a parent function and applying transformations to it.
Point Field		○	○			E	35	○	○	○	○	○	Students use a point field to investigate linear combinations in the form $ax + by$.
Build Your Own Dynagraph		○	○	○		C	35	○	○				Students or teachers build their own dynagraphs from scratch.

Legend: E = Easy; I = Intermediate; C = Challenging

© 2007 Key Curriculum Press

1

Functions

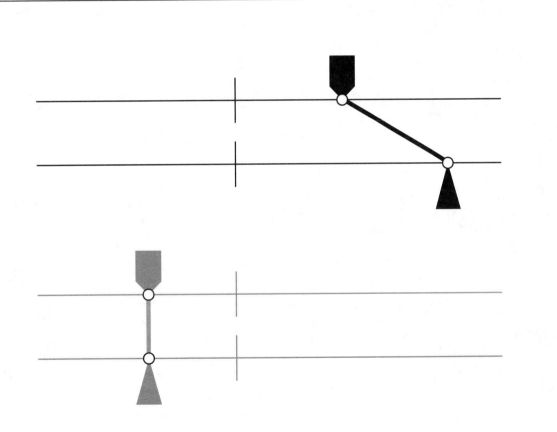

Introducing Dynagraphs

A *function* is a mechanism that gives you one specific output value for any value that you put in.

How many ways are there to represent a function? You've probably encountered various representations of functions, using tables, graphs, or equations. In this activity you'll explore a new way of representing functions: *dynagraphs*.

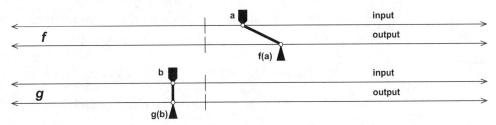

SKETCH AND INVESTIGATE

Q1 Spend a few moments reviewing with your group or on your own what a function is. Based on what you already know, how would you describe functions to someone who isn't familiar with them?

Each dynagraph has an input axis with an input marker and an output axis with an output marker. There is also a tick mark in the middle of each axis.

1. Open **Introducing Dynagraphs.gsp.** You'll see four dynagraphs labeled *f, g, h,* and *j,* each in a different color.

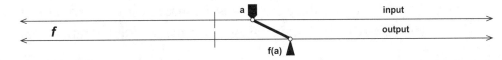

2. The input marker for dynagraph *f* is labeled *a.* To get an idea of how dynagraphs work, use the **Arrow** tool to drag this input marker.

Q2 Based on your understanding of functions, does this dynagraph represent a function? Explain.

Next you'll explore and describe in detail each of the dynagraphs on this page.

Here's a description of the *f* dynagraph:

This description is in terms of the position and motion of the input and output markers. The description does not use numbers or formulas because there are no numbers or formulas on the dynagraphs.

When the input marker is at the tick mark, the output marker is also at the tick mark. When the input marker is not at the tick mark, the output is always on the same side of the tick mark as the input. The output is always farther away from the tick mark than the input; it seems to be about twice as far away. When the input is dragged steadily from left to right, the output also moves steadily in the same direction, only faster.

Q3 Drag the input markers for dynagraphs *g, h,* and *j,* and then write detailed descriptions of these functions. Imagine you're describing the dynagraphs to someone who can't see them.

NUMBERS, NUMBERS, NUMBERS

You may have thought that it would be convenient to have number lines as the dynagraph axes to make it easier to give a precise description of the behavior of each function. With only a single tick mark, it's impossible to assign numbers to positions, such as "an input of 3 gives an output of 5."

In this section you'll explore four new dynagraphs, first without numbers, then with.

3. On page 2 are the new dynagraphs. Explore each by dragging its input marker.

Q4 Write a description of each function, just as in the previous section.

4. Press the *Show Number Lines* button. The dynagraph axes appear as number lines.

5. Drag *t*'s input control to 4.

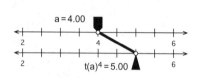

The arrow points to an output of 5, as shown here. Using function notation, you can write $t(4) = 5$, which is read "*t* of four equals five."

Q5 Solve for these unknowns. Be sure to use the correct function dynagraph for each one. Write each answer using function notation.

Hint: The answers to all but two of these questions are single numbers. One answer is "undefined," and one consists of several numbers.

a. $t(1) =$ b. $t(5) =$

c. $t(x) = -5, x =$ d. $u(-1) = g, g =$

e. $u(3) = p, p =$ f. $u(m) = 6, m =$

g. $v(4) =$ h. $v(-4) =$

i. $v(r) = 3, r =$ j. $w(2) = z, z =$

k. $w(4) = s, s =$ l. $w(a) = 0, a =$

EXPLORE MORE

A function has an *absolute maximum* if there is a largest output value—one the function can reach but can never exceed. Similarly, a function has an *absolute minimum* if there is a smallest output value—one the function can reach but can never go below.

Q6 Of the eight functions in the sketch, which functions have an absolute maximum or an absolute minimum? What are these maximum/minimum output values, and for what input values do they occur?

Exploring Algebra 2 with The Geometer's Sketchpad
© 2007 Key Curriculum Press

Introducing Dynagraphs

Objective: Students explore dynagraphs, an alternative to Cartesian graphs, to develop a feel for various types of functional relationships.

Student Audience: Algebra 1/Algebra 2

Prerequisites: Students should have some idea of what is meant by a function.

Sketchpad Level: Easy. Students manipulate a pre-made sketch.

Activity Time: 30–40 minutes. Be sure to give students enough time to write detailed and precise descriptions of the dynagraphs (Q3 and Q4). To reduce the amount of time required, students could skip Q4 or the Explore More section.

Setting: Paired/Individual Activity (use **Introducing Dynagraphs.gsp**) or Whole-Class Presentation (use **Introducing Dynagraphs Present.gsp**)

The term *dynagraph* was coined by Paul Goldenberg, Philip Lewis, and James O'Keefe in their study "Dynamic Representation and the Development of a Process Understanding of Functions" published by Education Development Center, Inc., and supported in part by a grant from the National Science Foundation.

The motivation for developing and using dynagraphs comes from the often-noted difficulty students have in seeing the graphs of functions as dynamic representations of functional relationships between two quantities and not just as static pictures. By decoupling the input and output axes, and having a segment connect points on parallel axes, students are better able to see the input-output machine view of functions expressed graphically. Being able to drag the input marker gives students the further advantage of actually varying the independent variable and seeing the function as a *dynamic* relationship between input and output.

Dynagraphs can serve as a bridge between the input-output machine model with which students are often introduced to functions and function graphs in the Cartesian plane.

SKETCH AND INVESTIGATE

Q1 Answers will vary, but should basically describe functions as consistent input-output machines. In other words, they are relations or mappings between input values and output values such that any valid input value maps to a single output value.

Q2 The dynagraphs do represent functions because they map input values to output values and they are consistent—a particular input value will always point to the same output value.

Q3 Answers will vary, but should not involve numbers or formulas. Good answers will in general include dynamic descriptions ("As the input is dragged steadily from left to right, the output . . .") and note any symmetries present.

NUMBERS, NUMBERS, NUMBERS

Q4 See Q3.

Q5
a. $t(1) = -1$ b. $t(5) = 7$
c. $x = -1$ d. $g = -6$
e. $p = -6$ f. $m = \ldots, -7, -3, 1, 5, \ldots$
g. $v(4) = 2$ h. $v(-4)$ is undefined
i. $r = 9$ j. $z = 1$
k. $s = 3$ l. $a = 2.5$

EXPLORE MORE

Q6 Function j has an absolute minimum of 0 at 0. Function u has an absolute maximum of 6 at $(\ldots, -7, -3, 1, 5, \ldots)$, and has an absolute minimum of -6 at $(\ldots, -5, -1, 3, 7, \ldots)$. Function v has an absolute minimum of 0 at 0.

WHOLE-CLASS PRESENTATION

Use **Introducing Dynagraphs Present.gsp** to explore this highly dynamic visual representation of functions with your students. Dynagraphs differ from Cartesian graphs in that you can make the variables really vary, so emphasize the variation in the presentation by using the Animation buttons and leaving variables moving on the screen.

Introducing Dynagraphs

This whole-class presentation allows students to gain a dynamic perspective on the notion of function and emphasizes the way in which the variables really vary.

SKETCH AND INVESTIGATE

Q1 Begin by asking students to describe a function in their own words. Get responses from several students, and encourage a diversity of descriptions. Consider forming small groups of two or three students and asking each group to create its own written description, suitable for explaining functions to someone who isn't familiar with them.

1. Open **Introducing Dynagraphs Present.gsp.** Four dynagraphs appear, each in a different color.

2. Explain that the input and output markers represent the variables and that this model allows you to vary the variables by dragging the markers. Use the **Arrow** tool to drag input marker *a*. After dragging it a bit, use the *Animate a* button to leave it in motion.

Q2 Ask students whether the behavior they observe represents a function, based on their description of what a function is. Solicit different explanations from as many students as possible.

Q3 Ask students to describe the behavior of this first function. They will want to call the tick mark "zero" or "the origin," and they will want to describe movement to the left or right as "increasing" or "decreasing." These characterizations are based on numbers; resist them, and instead encourage students to describe the behavior in terms of position, movement, and symmetry. Consider asking students whether the function has a "fixed point"—a state in which the input marker and the output marker are at exactly the same position.

Q4 Drag the input marker for the second function and have students observe. Leave it in motion while students describe the behavior. (Students often want to call this function a "constant function." Rather than describing this answer as wrong, ask them whether it's the output that is constant or whether there is something else about this function that they view as being "constant.")

Leave each function in motion while you drag and discuss the remaining functions.

Q5 Use the *Animate c* button to put the third function's input marker into motion. (Students will often laugh at this function, and you may want to ask them how often they have laughed at a mathematical function.) Have them describe this function in detail. They may want to give it a name.

3. The remaining pages show dynagraphs with numbers added to the axes. Have students answer the questions on each page, and then allow them to see the algebraic formulas underlying the behavior of the dynagraphs.

From Dynagraphs to Cartesian Graphs

Dynagraphs make it easy to change the input of a function and see how each input produces a corresponding output. This strength is also a weakness, because you can see only a single pair of input-output values at any time.

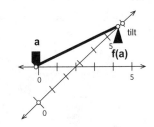

In this activity you'll change a dynagraph so that you can keep track of many input and output values at the same time.

FUNCTION MATCHMAKING

Start out with some "function matchmaking"—you'll match several dynagraphs to their corresponding algebraic equations.

The dynagraphs don't show any numbers, so you'll have to figure out each match by observing the output marker as you drag the input marker.

Q1 Open **Dyna To Cartesian.gsp.** The five dynagraphs on page 1 (labeled "Matching") correspond to the equations below. Pair each dynagraph with an equation and explain how you made the match.

a. $y = x$
b. $y = -x$
c. $y = 2x$

d. $y = x^2$
e. $y = \dfrac{1}{x-1} + 1$

FROM DYNAGRAPHS TO CARTESIAN GRAPHS

1. On page 2 of **Dyna To Cartesian.gsp,** you'll see a dynagraph for the function $f(x) = 2x - 1$. Drag the input marker to familiarize yourself with this function.

2. Drag the point labeled *tilt* so that the output axis is at an angle to the input axis. Drag the input marker again. Do this for a few different angles of the output axis. (You can even turn it upside down!)

Q2 When you tilt the output axis, what changes and what remains the same?

3. Press the *Make Cartesian* button and watch as the two axes of your dynagraph "morph" into the familiar *x*- and *y*-axes. Drag the input once more to convince yourself that you're still dealing with the same dynagraph, only tilted.

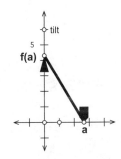

4. Press the *Show Perpendiculars* button to show lines through *a* and *f(a)* perpendicular to the two axes.

Click on the intersection to construct the point of intersection.

5. Construct the intersection of the two perpendiculars.

6. With the new point selected, measure its coordinates by choosing **Measure | Coordinates.**

Q3 Drag the input marker. What does the *x*-coordinate of the new point correspond to on the dynagraph? What does the *y*-coordinate correspond to?

7. Deselect all objects by clicking in blank space. Select the new point and choose **Display | Trace Intersection**. Now drag or animate the input marker and watch as P traces out the graph of $f(x)$.

Q4 Describe the shape of the graph traced by the intersection point. Why does this shape make sense given the behavior of the dynagraph?

SIMULTANEOUS REPRESENTATION

8. Go to page 3 of **Dyna To Cartesian.gsp**.

You'll see a dynagraph and a Cartesian graph, both modeling $f(x) = 2x - 1$. Drag the input marker on the dynagraph and watch both models change simultaneously.

Q5 A classmate says, "One Cartesian point contains the same information as two dynagraph points." Explain what she means.

Double-click the function equation f(x) = 2x − 1 to edit it.

9. Explore each of the following functions on the combined dynagraph/Cartesian graph. Enter the function, then drag the input marker slowly from left to right. Observe what happens to the point on the Cartesian graph as you drag.

$$f(x) = 3 \qquad f(x) = x \qquad f(x) = -x$$
$$f(x) = x^2 \qquad f(x) = -x^2 \qquad f(x) = 5x$$

Q6 Fill in the blanks.

When the input and output markers both move right, the Cartesian point moves _____.

When the input marker moves right and the output marker moves left, the Cartesian point moves _____.

Q7 How does the Cartesian graph of $f(x) = 5x$ compare to that of $f(x) = x$? How does this relate to the difference between their dynagraphs?

Q8 Compare dynagraphs and Cartesian graphs. In what ways do you think dynagraphs are better for representing functions? In what ways do you think Cartesian graphs are better?

From Dynagraphs to Cartesian Graphs

Objective: Students make connections between symbolic and dynagraph representations of functions by matching dynagraphs with equations, and then generate a Cartesian representation by geometrically manipulating a dynagraph.

Student Audience: Algebra 1/Algebra 2

Prerequisites: Students should already have been introduced to dynagraphs.

Sketchpad Level: Easy. Students manipulate a pre-made sketch and add to it minimally.

Activity Time: 40–50 minutes

Setting: Paired/Individual Activity (use **Dyna To Cartesian.gsp**) or Whole-Class Presentation (use **Dyna To Cartesian Present.gsp**)

The term *dynagraph* was coined by Paul Goldenberg, Philip Lewis, and James O'Keefe in their study "Dynamic Representation and the Development of a Process Understanding of Functions" published by Education Development Center, Inc., and supported in part by a grant from the National Science Foundation.

FUNCTION MATCHMAKING

Q1 Explanations will vary, but should involve the relative direction and speed of the input and output. Students can animate the input marker to make it easier to judge relative speed. Explanations may also refer to range restrictions (in d) and the existence of an asymptote (in e). Here are the matches:

a. 4   b. 2   c. 5

d. 1   e. 3

FROM DYNAGRAPHS TO CARTESIAN GRAPHS

Q2 Answers will vary. When the output axis is tilted, the two axes intersect, and there's a change in the length and orientation of the segment connecting the two dynagraph markers. Even as these things change, the output value remains the same for any given input value, so the two markers continue to represent the same function. If $f(a) = b$ before tilting, $f(a) = b$ after tilting.

4. Consider having students construct the perpendiculars on their own instead of pressing the button.

Q3 The x-coordinate of the Cartesian point is a (the input value of the dynagraph). Similarly, the y-coordinate of the new Cartesian point is $f(a)$ (the output value of the dynagraph).

7. After generating the trace, students can use **Construct | Locus** to create a permanent image. But it's important that they trace first, to get a visual image of the way in which the dynagraph traces out the Cartesian graph over time.

Q4 The graph is a line with a positive slope. The fact that it's a line corresponds to the fact that as the input marker on the dynagraph is dragged at a constant speed, the output marker moves at a constant speed. Students may make additional observations: How far the input marker goes in a given unit of time corresponds to *run*, and how far the output marker goes corresponds to *rise*. The fact that the line has a positive slope corresponds to the fact that the two markers always go the same direction (see Q6). The fact that the slope of the line is greater than 1 corresponds to the fact that the output marker moves faster than the input marker. (A slope of 1 corresponds to markers moving at the same speed; a slope between 0 and 1 corresponds to a slower output marker.)

SIMULTANEOUS REPRESENTATION

Q5 A dynagraph point contains one piece of information: its position on its axis. Thus two dynagraph points are required to show both input and output values for a function. A Cartesian point contains two pieces of information—its x-coordinate and its y-coordinate—so the Cartesian point can represent both the input and the output values at the same time.

Q6 When the input and output markers both move right, the Cartesian point moves up and to the right.

When the input marker moves right and the output marker moves left, the Cartesian point moves down and to the right.

Q7 The Cartesian graph of $f(x) = 5x$ is *steeper* than the graph of $f(x) = x$. This corresponds to the fact that the output marker moves faster for the dynagraph of $f(x) = 5x$ than it does for that of $f(x) = x$.

Q8 Student answers will vary, but should describe one advantage of each representation. Students may say that it's easier to distinguish input values from output values on a dynagraph, and that the dynagraph allows them to actually vary the variable. The dynagraph relates more closely to the fundamental definition of a function: For any given input value, it produces one single output value. The Cartesian graph makes the global picture more accessible: At a glance you can see the characteristics of the entire function for all input values, with no need to drag or animate anything. Finally, the dynagraph shows clearly where each point on the Cartesian graph comes from.

WHOLE-CLASS PRESENTATION

In this presentation, students make connections between the behavior of a dynagraph that represents a function and an equation representing the same function. They then see how modifying the dynagraph representation produces a Cartesian representation, and they view different functions in both representations simultaneously.

1. Open **Dyna To Cartesian Present.gsp.** Press *Animate* to put the top dynagraph into motion.

Q1 Have students guess which of the five equations on the bottom of the screen corresponds to the top dynagraph. Ask them to give arguments for their guesses, and encourage observations that relate the direction or speed of the dynagraph markers to the equations. Repeat for all five functions.

2. Page 2 contains a single dynagraph. Put it in motion.

Q2 Ask students what they think will happen if you drag the point labeled *tilt*.

Q3 With the input marker still moving, drag point *tilt* so that the bottom axis is only slightly off horizontal. Ask students to observe what has changed about the behavior of the dynagraph and what has stayed the same. Try this with several different positions of point *tilt*. The class should agree that although the direction of the output marker's motion has changed, the motion of the marker along its axis, and its numeric value, remain unchanged.

3. With the input marker in motion, press the *Make Cartesian* button.

Q4 Ask students to report what they observe about the angle of the output axis and about the origin of the output axis.

4. With the input marker in motion, press the *Show Perpendiculars* button and ask students to observe the relationship between the dashed lines and the values of a and $f(a)$.

5. Show the intersection.

Q5 Ask students to observe the path of the intersection: its shape, its location on the screen, and so forth.

6. Turn tracing on so that students can check their answers.

7. Consider using this page to look at one or two different functions, such as $f(x) = x^2$ or $f(x) = |x|$.

8. Page 3 contains both dynagraph and Cartesian representations. Animate the input marker a.

Q6 Ask students what relationships they can detect about the connections between the two models. Students may make observations about the behavior of the markers on the four axes, or about the relation between the moving point labeled $(a, f(a))$ and the motion of the dynagraph.

9. Edit the equation to change the function being displayed, and ask students how looking at a different function changes their observations about the connections.

Finish with these three questions:

Q7 Ask students to fill in the blanks:

When the input and output markers both move right, the Cartesian point moves _____.

When the input marker moves right and the output marker moves left, the Cartesian point moves

_____.

Q8 How does the Cartesian graph of $f(x) = 5x$ compare to that of $f(x) = x$? How does this relate to the difference between their dynagraphs?

Q9 Compare dynagraphs and Cartesian graphs. In what ways do you think dynagraphs are better for representing functions? In what ways do you think Cartesian graphs are better?

Domain and Range

You can't put a television in a blender, and you wouldn't expect an elephant to come out of a gasoline pump. In math terms, a television isn't an *allowable input* for a blender; it's not part of a blender's *domain*. And an elephant isn't a *possible output* of a gasoline pump; it's not part of a gas pump's *range*.

Similarly, functions have certain numbers that are and aren't allowed as inputs, and other numbers that are and aren't possible as outputs. In this activity you'll explore these notions using both dynagraphs and Cartesian graphs.

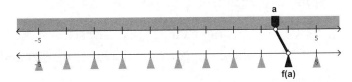

SKETCH AND INVESTIGATE

The *range* of a function is the set of possible outputs from that function. Let's see how dynagraphs can make this idea clearer. You'll start by exploring the range of everyone's favorite dynagraph: the "blue hopper."

1. Open **Domain Range.gsp.** Drag the input marker to observe the behavior of the function $f(x) = \text{round}(x)$. The input marker leaves a trace.

You can drag the input marker to any value you want, so we say "the domain of f is all real numbers."

To turn on tracing, select the output marker and choose **Display | Trace Triangle.**

2. Turn on tracing for the output marker. Then drag the input marker back and forth again.

The output marker leaves a trace of where it's been. These traces point to every integer but never to any other values, so we say "the range of f is all integers."

The range of each function on page 2 will either be "all real numbers" or an inequality such as $f(x) \geq 5$.

Q1 On page 2, use the same method to find the range of each of the four functions.

When you use technology, it's very important to think about the limitations of that technology. You'll see that the method used above can be misleading in certain situations, and you'll then learn a more reliable method.

3. On page 3, turn on tracing for the output marker, then drag the input marker, as in step 2. For greater control, use the right and left arrow keys on your keyboard to drag one pixel (screen unit) at a time.

Q2 What does the range of $k(x) = 20 \cdot x$ *appear* to be? Explain why this answer is actually wrong. Why do you think this happens?

4. Select the input marker and choose **Display | Animate Pentagon.** Repeatedly press the *Decrease Speed* button (the down arrow on the Motion Controller) until it's clear that the range of this function really is all real numbers. Go back to pages 1 and 2 of the sketch, and convince yourself that your answers there were correct.

For most functions, the domain is "all real numbers," meaning that any input produces some output. Sometimes, however, a domain might be *restricted* to something such as "all integers" or "$x > 3$."

Q3 On page 4, drag v's input marker back and forth. What is v's domain? (In other words, where can you drag the input marker and still see the output marker?) What is v's range?

Q4 Based on its equation, why are some numbers not part of v's domain?

Q5 The domain of w is all real numbers except for one particular value. The range of w is also all real numbers except for one particular value (a different value). What is the one value not in the domain of w? What is the one value not in the range of w?

DOMAIN AND RANGE ON CARTESIAN GRAPHS

Let's transfer this knowledge to Cartesian graphs.

5. Page 5 shows a rounding function. Drag the input marker and think about how you can tell domain and range on a Cartesian graph.

To edit the function, double-click its equation and then enter the new expression.

6. On page 6, explore the function $f(x) = 2x$. Then edit the function and explore the functions listed here, again thinking about how to tell domain and range on a Cartesian graph.

$$g(x) = \frac{x}{2} \qquad h(x) = x^2 \qquad j(x) = 2x + 1 \qquad k(x) = 2\sin\left(\frac{\pi x}{2}\right)$$

Q6 How can you tell the range of a function just by looking at its Cartesian graph? How about its domain?

EXPLORE MORE

Q7 On page 1, change f's equation so that its range is all even numbers. Then change it again to make it all odd numbers. Record the equations you used.

Q8 On page 4, change v so that both its domain and its range include only numbers less than or equal to 0. Change w so that its domain is all real numbers except 0 and its range is all real numbers except 2. Record the equations you used.

Domain and Range

Objective: Students explore the domain and range of various functions using dynagraphs and Cartesian graphs. They drag input variables and observe the behavior of the input and output of the functions, with special attention to functions having restricted domain and/or range. Finally, they design functions to have particular domain and range restrictions.

Student Audience: Algebra 1/Algebra 2

Prerequisites: Students should already have been introduced to dynagraphs in a previous activity such as Introducing Dynagraphs. Some prior introduction to domain and range is also very helpful for most students.

Sketchpad Level: Easy. Students manipulate a pre-made sketch and modify it only minimally.

Activity Time: 30–40 minutes

Setting: Paired/Individual Activity (use **Domain Range. gsp**) or Whole-Class Presentation (use **Domain Range Present.gsp**)

Related Activities: Introducing Dynagraphs, From Dynagraphs to Cartesian Graphs, Function Composition with Dynagraphs

The term *dynagraph* was coined by Paul Goldenberg, Philip Lewis, and James O'Keefe in their study "Dynamic Representation and the Development of a Process Understanding of Functions" published by Education Development Center, Inc., and supported in part by a grant from the National Science Foundation. See the Activity Notes from the activity Introducing Dynagraphs for a more thorough discussion of dynagraphs in general.

It's a good idea to have an initial class discussion about domain and range before starting the activity, and it's very important to have a class discussion after students finish the activity.

One or the other of these discussions should focus on why some numbers are not allowed as inputs for particular functions. Ask students to give examples of functions with restricted domains. One category of answers is covered in the activity—functions such as square root functions and rational functions that have undefined outputs for certain inputs. Another category, not covered in the activity, is functions that model real-life or geometric situations.

For example, the function $f(n) = 0.89n$ might represent the total cost of buying n apples each costing $0.89. But it doesn't make sense to consider $n = -2$ or $n = 3.71$ here. Only positive whole numbers are part of this domain.

A good way to encourage exploration is to ask students to modify the various functions and observe the results. On page 1, you could ask students how the range would change if they changed the function to $f(x) = \text{round}(x) + 1$ (*answer:* it wouldn't). You could then ask for what values of k would the function $f(x) = \text{round}(x) + k$ have a range different from that of the original function (*answer:* non-integer values). You could follow this up by exploring what happens when you multiply $\text{round}(x)$ by a constant. You could extend the activity in similar ways on the other pages.

SKETCH AND INVESTIGATE

Q1 $g(x)$: all real numbers
$h(x)$: $h(x) \geq 0$
$j(x)$: all real numbers
$k(x)$: $-2 \leq k(x) \leq 2$

Q2 The range appears to be all integers, as with $f(x) = \text{round}(x)$. This is clearly wrong since many inputs result in non-integer outputs. For example, $g(0.01) = 0.2$.

The output marker lands only on integers because the input marker can't really be dragged continuously. You can only move it by one pixel (screen unit) at a time. This dynagraph's scale is set to 1 pixel = 0.05 units. Moving the input marker by a single pixel moves the output marker by 20 times as much, which is equivalent to a full unit. You can change the scale by pressing the *Show Scale* button and adjusting the slider.

In the next step, students are encouraged to use animation rather than dragging the marker directly, because an object being animated can be slowed down so that it moves less than a pixel at a time.

Q3 The domain of function $v(a)$ is $a \geq 0$. (When the input marker is to the left of 0, the output marker disappears.)

The range is $v(a) \geq 0$.

Q4 Negative numbers are not in the domain because the square root of any negative number is undefined over the set of real numbers.

Q5 The domain restriction is $b \neq 2$, and the range restriction is $w(b) \neq 0$. The value 2 is excluded from the domain because it would result in division by 0. The value 0 isn't part of the range because the result of the division $\frac{1}{x-2}$ cannot be 0.

DOMAIN AND RANGE ON CARTESIAN GRAPHS

On page 5, the graph's properties are set to plot it discretely rather than continuously. If the plot were done continuously, the discontinuities would be connected with segments. To change whether a graph is plotted continuously or discretely, select the object and choose **Edit | Properties | Plot.**

Q6 One way to determine domain by looking at a Cartesian graph of a function is to imagine a vertical line that sweeps from left to right. Any location where the line touches the graph is part of the domain. Thus, if a vertical line crossing the x-axis at $x = 3$ touches the graph somewhere, 3 is part of the domain of that function. Any location where the line doesn't touch the graph at all is not part of the domain.

Similarly, to determine the range from a Cartesian graph, imagine a horizontal line sweeping from bottom to top. Anywhere it touches some part of the graph is part of the range; anywhere it doesn't, isn't.

EXPLORE MORE

Q7 To make the range of f all even numbers, use $f(x) = 2 \cdot \text{round}(x)$.

To make its range all odd numbers, use $f(x) = 2 \cdot \text{round}(x) + 1$.

Q8 To make the domain and range of v all numbers less than or equal to 0, use $v(x) = -\sqrt{-x}$.

To make w's domain all real numbers except 0 and its range all real numbers except 2, use $w(x) = \frac{1}{x} + 2$.

WHOLE-CLASS PRESENTATION

Open **Domain Range Present.gsp** and use the pages of this sketch to stimulate a class discussion.

1. On page 1, drag input marker a and ask students to observe the possible positions of the input and output markers, and use their observations to describe the domain and range. After several students have volunteered descriptions in their own words, turn on tracing and drag again to verify the descriptions. You can use the Animation button to achieve smooth movement of the input marker.

2. For each of the four functions on page 2, ask students to guess ahead of time what the range will be. Then drag the input marker for that function to test their guesses. For each function, use tracing and animation to generate a smooth, detailed visual representation of the answer.

3. Use page 3 to emphasize the need to pay attention to details and not to jump to conclusions. By dragging the input marker, you'll generate what appears to be a range of integer values only. Ask students to explain what's going on here. This should generate a lively discussion. Encourage a number of students to describe the phenomenon in their own words. Finish this page by using animation to achieve movement by less than a pixel at a time.

4. On page 4, tracing shows the domain and range restrictions for $v(a)$ clearly, but cannot show the restrictions for $w(b)$ so clearly. Get students to discuss the differences in the two situations, so that they realize that $w(b)$ is missing only a single number in its domain and a different number in its range.

5. On page 5, drag input marker a and have students observe both the dynagraph and Cartesian graph. Ask them to explain how the restricted range shows up on the Cartesian graph. Also ask them how to edit the function to generate only even numbers, or only odd numbers.

6. Use page 6 to compare the dynagraph and Cartesian representations for the functions from page 2.

Finish by asking students to summarize what they learned about domain and range.

Function Composition with Dynagraphs

In life, the answer to one question sometimes becomes a question that leads to another answer. Functions are much the same; sometimes we take the output of one function and make it the input for a second function. This is called *function composition*, and we say that the two functions have been *composed*. In this activity you'll get a brief introduction to function composition and then see how dynagraphs can provide a compelling way of modeling composed functions.

INTRODUCTION

We'll introduce function composition informally by doing some examples with numbers.

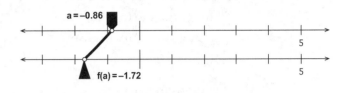

The composite function $g(f(3))$ is pronounced "g of f of 3."

Given $f(x) = 2x$ and $g(x) = x^2$, find $g(f(3))$.

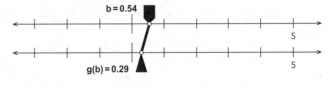

You always evaluate parentheses first, so start on the inside by evaluating $f(3)$: $f(3) = 2 \cdot (3) = 6$.

Take this output and make it g's input: $g(f(3)) = g(6) = (6)^2 = 36$.

That, in a tiny nutshell, is function composition.

Q1 Given the functions

$$f(x) = 2x, g(x) = x^2, h(x) = \text{round}(x), \text{ and } j(x) = \frac{x}{2},$$

evaluate the following expressions:

a. $g(f(5))$ b. $f(g(3))$ c. $f(h(3.6))$

d. $j(g(-6))$ e. $j(f(17))$ f. $f(j(17))$

Q2 Do you think $f(g(x))$ always equals $g(f(x))$? Answer this question by comparing $f(g(5))$ and $g(f(5))$.

SKETCH AND INVESTIGATE

1. Open **Composite Functions.gsp.** Drag each input marker to familiarize yourself with the dynagraphs of functions f and g.

You'll now model $g(f(x))$. The trick is to use Sketchpad's **Split** and **Merge** commands so that the output of f becomes the input of g.

2. Select point b (at the tip of g's input marker) and choose **Edit | Split Point From Line.** The point is separated from its axis.

3. Select points b and $f(a)$. Choose **Edit | Merge Points.**

The two points are merged into one, and the output of f is the input of g. Drag the input marker of f to explore your new composite function, $g(f(x))$, and to check your answer from Q1 part a.

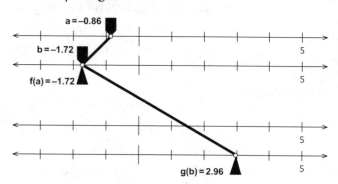

4. Go to page 2, which contains the same two functions, but with a g above and f below. Use the technique from steps 2 and 3 to model $f(g(x))$.

Q3 Use your composite dynagraph to evaluate these expressions:

    a. $f(g(1))$ b. $f(g(-1))$ c. $f(g(-7))$

Next, you'll compose the "round" and "square" functions to create a composite function with an interesting set of outputs.

Q4 Go to page "h&g" and model $g(h(x))$. What is the range of this composite function? In other words, what are its possible outputs?

Use the remaining pages of the document to experiment with other composite functions.

Q5 Create the composite functions $j(f(x))$ and $f(j(x))$ on the appropriate pages. Experiment with these functions. What special feature do you notice about these two composite functions? Why does this happen?

EXPLORE MORE

1. There's nothing stopping you from composing more than two functions to get something such as $h(g(f(x)))$. Go to page "all 4" and try this for different combinations of three or four functions. See if you can build the following functions:

    a. a function that outputs twice perfect squares $(0, 2, 8, 18, \ldots)$

    b. a function that outputs squares of even numbers $(0, 4, 16, 36, \ldots)$

    c. a function that outputs the perfect squares divided by 4 $(0, 0.25, 1, 2.25, 4, 6.25, \ldots)$

Exploring Algebra 2 with The Geometer's Sketchpad
© 2007 Key Curriculum Press

Function Composition with Dynagraphs

Objective: Students use dynagraphs to model composite functions, using the output of one function as the input to a second function.

Student Audience: Algebra 2/Precalculus

Prerequisites: It's helpful if students have already been introduced to dynagraphs by doing the first dynagraph activity: Introducing Dynagraphs. They should also be familiar with function notation.

Sketchpad Level: Easy. Students manipulate a pre-made sketch and add on to it minimally.

Activity Time: 35–45 minutes

Setting: Paired/Individual Activity (use **Composite Functions.gsp**) or Whole-Class Presentation (use **Composite Functions Present.gsp**)

Related Activities: Introducing Dynagraphs, From Dynagraphs to Cartesian Graphs, Domain and Range

The term *dynagraph* was coined by Paul Goldenberg, Philip Lewis, and James O'Keefe in their study "Dynamic Representation and the Development of a Process Understanding of Functions" published by Education Development Center, Inc., and supported in part by a grant from the National Science Foundation. See the Activity Notes from the activity Introducing Dynagraphs for a more thorough discussion of dynagraphs in general.

Composite functions often seem backward to students. Since they read $g(f(x))$ left to right, they may be surprised to see that its composite dynagraph starts at the input marker of f and ends at the output marker of g. The reason, of course, is that when evaluating $g(f(x))$, you start with the parentheses and apply the functions from right to left, even though you read the expression from left to right. Make sure students really do understand why composite dynagraphs are structured as they are.

A possible discussion topic with students is the fact that function composition lurks behind *every* function. For example, the simple function $f(x) = 2x + 1$ can be seen as $g(h(x))$, where $g(x) = x + 1$ and $h(x) = 2x$, or $g(x) = 2x$ and $h(x) = x + 0.5$, or $g(x) = 2x + 1$ and $h(x) = x$, or

many other possibilities. Students may enjoy decomposing functions in this way, and the practice will be valuable when they learn to transform functions.

INTRODUCTION

Q1 a. 100   b. 18   c. 8
       d. 18   e. 17   f. 17

Q2 No, $f(g(x))$ doesn't always equal $g(f(x))$. In this case, $f(g(5)) = 50$ and $g(f(5)) = 100$.

SKETCH AND INVESTIGATE

4. The only way to undo merging an input point with an output point is literally to undo. That is, choose **Edit | Undo** repeatedly (or hold down the Shift key and choose **Edit | Undo All**). Splitting (by choosing **Edit | Split**) won't work because the original output points were constructed in a way that does not allow them to be split.

Q3 a. 2   b. 2   c. 98

Q4 The range of $g(h(x))$ is all squares of whole numbers, or, (0, 1, 4, 9, 16, 25, 36, . . .).

A nice way to see this is to turn on tracing for the output marker (by selecting it and choosing **Display | Trace Triangle**) and then drag the input marker. This is, in fact, the method used in the activity Domain and Range.

Q5 The output is always the same as the input in both cases. This is because $f(x)$ and $j(x)$ are inverse functions of each other, meaning that they "undo" each other. In other words, start with any number, double it, then halve it, and you'll end up where you started. It works the same if you halve it first, then double it.

EXPLORE MORE

1. a. $f(g(h(x)))$
    b. $g(f(h(x)))$
    c. $g(j(h(x)))$

By representing composite functions on a dynagraph, it is possible to show a number graphically as it leaves one function and becomes the argument of a second function. Seeing the concept in another way may help students to understand the process as opposed to memorizing a rule.

1. Open **Composite Functions Present.gsp.** The first page shows two functions: $f(x) = 2x$ and $g(x) = x^2$. On each function, drag the input marker (the pentagon above the axis) and tell the class to observe the output.

Q1 What is the range of each function? (The range of f is all real numbers. The range of g is all non-negative real numbers.)

Explain that function composition means using the output from f as the input to g.

Q2 Drag the input marker for g so its value is reasonably close to the output for f. Ask if this is close enough. Get students to agree that you need a better method.

2. Select point b, the input of g. Choose **Edit | Split Point From Line.**

3. Select points b and $f(a)$. Choose **Edit | Merge Points.**

Q3 What we have done graphically is to set b equal to $f(a)$. Write "$b = f(a)$" in the sketch or on the board. Point to the marker for $g(b)$ on the sketch. Using substitution, what would be a logical name for this value? (Pronounce it slowly as you write "g of f of a.")

$$g(b) = g(f(a))$$

4. The next page is labeled "g&f". Use the same procedure to construct $f(g(b))$. Flip between the two pages as you compare the composite functions.

Q4 What are the ranges of $g(f(x))$ and $f(g(x))$? (All non-negative real numbers in both cases.)

Q5 Is $g(f(x)) = f(g(x))$ for all x? This is true only for $x = 0$, but challenge students to work it out themselves before showing a counter-example.

5. Go to the page labeled "f&j", which has the functions $f(x) = 2x$ and $j(x) = \frac{x}{2}$.

Q6 What are $f(j(x))$ and $j(f(x))$? (They are both the identity x.)

6. Perform the constructions to answer Q5.

You can undo all of the modifications to any page in the document. While holding down the Shift key, choose **Edit | Undo All.**

In this document, there are a total of four functions. They all appear on the last page, "all 4". Try the various combinations. In each case, ask students about the range of the composite function, and ask them whether the order of the functions matters.

You can also customize these functions. Double-click on a function definition on the left side of the screen, and enter a new definition.

Odd and Even Functions

Just as there are odd and even numbers, there are odd and even functions. Unlike numbers, there are also functions that are neither odd nor even. Whether a function is odd, even, or neither depends on its symmetry. In this activity you'll explore odd and even functions using both dynagraphs and Cartesian graphs.

SKETCH AND INVESTIGATE

1. Open **Odd Even Functions.gsp.**

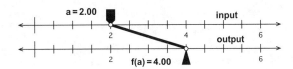

Q1 Use the first dynagraph to determine these values:

   a. $f(2)$ and $f(-2)$ b. $f(-1)$ and $f(1)$

   c. $f(3)$ and $f(-3)$ d. $f(-0.5)$ and $f(0.5)$

Q2 In each case, what do you notice about how $f(a)$ compares with $f(-a)$?

Functions like this, where the output for $-x$ is the *opposite* of the output for x, are said to be *odd* or to have *odd symmetry*. We can say the following:

An *odd function* is one in which $f(-x) = -f(x)$ for all x in its domain.

Q3 Of the other three functions modeled on page 1, which are odd?

Q4 On page 2, use the first dynagraph to determine these values:

   a. $t(2)$ and $t(-2)$ b. $t(-1)$ and $t(1)$

   c. $t(3)$ and $t(-3)$ d. $t(-0.5)$ and $t(0.5)$

Q5 In each case, what do you notice about how $t(x)$ compares with $t(-x)$?

Functions like this, where the output for $-x$ is the *same* as the output for x, are said to be *even* or to have *even symmetry*. We can say the following:

An *even function* is one in which $f(-x) = f(x)$ for all x in its domain.

Q6 Of the other three functions modeled on page 2, which are even?

Q7 Describe how to tell whether a dynagraph you're exploring represents an odd or an even function just by dragging (in other words, without looking at its equation).

Q8 Go to page 3. Model the three odd functions from page 1 and the three even functions from page 2 on the combination dynagraph/Cartesian graph. What do odd functions look like when plotted in the xy plane? How about even functions?

To enter round(x), |x|, sin(x) or cos(x) in the Edit Function dialog box, use the Functions pop-up menu. Don't worry if you're not familiar with the sine or cosine functions—you can tell whether they're odd or even just as with the other functions.

Q9 Model the following functions on the combination dynagraph/Cartesian graph. Some of the functions are odd, some are even, and one is neither. Use what you learned in Q8 to determine which are which, without using the dynagraph.

a. $f(x) = 5x$

b. $f(x) = x^3 - 2x$

c. $f(x) = \sin(x)$

d. $f(x) = x^2 + 2$

e. $f(x) = x^4 - 3x^3$

f. $f(x) = \cos(x)$

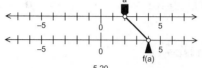

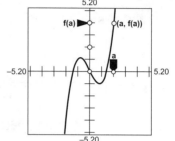

EXPLORE MORE

On page 4, you will construct transformed images of a graph and compare odd and even functions with the transformed images of their graphs.

To reflect *P*, select the x-axis and choose **Transform | Mark Mirror.** Then select *P* and choose **Transform | Reflect.**

To construct the locus, select both *P* and its reflected image, and choose **Construct | Locus.**

2. Page 4 contains the graph of a polynomial function and a point *P* on the graph. Reflect *P* across the *y*-axis. Then construct the locus of the reflected point as *P* moves along the graph. Make the locus thick and light blue.

3. Construct another transformed image of the original graph, this time reflected across the *x*-axis. Construct a third transformed image of the original graph, this time reflecting *P* across first the *y*-axis and then the *x*-axis. Make each locus thick and a different light color from the others. Record your observations.

Q10 Edit the original function to match each of the functions listed in Q9. For each function, record your observations about the original function and the images.

Q11 Why are odd functions called "odd" and even functions called "even"? You may be able to figure this out by looking at the equations of the odd and even functions you've seen in this activity. (*Hint:* Focus on just the polynomial equations.) Test your answer by modeling your own odd and even functions in Sketchpad and verifying that they exhibit the characteristic symmetry.

Q12 Imagine that both $f(x)$ and $g(x)$ are odd functions. What kind of function will $(f + g)(x)$ be? What about $(f \cdot g)(x)$? What about $g(f(x))$? What will the results be if f and g are both even? What if one is even and the other odd?

Odd and Even Functions

Objective: Students explore odd and even functions using dynagraphs, identify characteristics of these functions in the Cartesian plane, and compare the original functions to transformed images of their graphs.

Student Audience: Algebra 2/Precalculus

Prerequisites: Students should already be familiar with dynagraphs by having completed the activities Introducing Dynagraphs and From Dynagraphs to Cartesian Graphs.

Sketchpad Level: Easy. Students manipulate a pre-made sketch by dragging markers and editing functions.

Activity Time: 30–40 minutes

Setting: Paired/Individual Activity (use **Odd Even Functions.gsp**) or Whole-Class Presentation (use **Odd Even Functions Present.gsp**)

Before introducing odd and even functions, tell students that mathematicians often classify mathematical objects based on properties that make them unusual or special. For example, even numbers are special because they're always divisible by 2; prime numbers are special because they have no proper divisors. What properties might make a function special? Students have already seen some: The round(x) function is unusual because it contains breaks—you can't draw it without lifting your pencil. Such functions are called *discontinuous*. This discussion will set the stage for the investigation of odd and even functions.

One important thing when doing this activity is the difference between "negative" and "opposite." In the equation $f(-x) = -f(x)$ that characterizes odd functions, the input on the left side of the equation isn't necessarily negative—it's just the opposite of the input on the right side. Similarly, the output on the right side isn't necessarily negative—it's just the opposite of the output on the left side. It's important that students understand this point.

In Explore More steps 2 and 3 and Q10, students actually construct the reflected functions. This helps to make their conclusions about the symmetry of the graphs more concrete and more obvious to them. If your students are already comfortable with Sketchpad constructions, you should make this section of the activity mandatory for them. If not, consider using the presentation sketch to show these transformed images and explore their relationship with even and odd functions.

SKETCH AND INVESTIGATE

Q1 a. $f(2) = 4; f(-2) = -4$

b. $f(-1) = 1; f(1) = -1$

c. $f(3) = 21; f(-3) = -21$

d. $f(-0.5) = 0.88; f(0.5) = -0.88$

Q2 $f(a)$ and $f(-a)$ are opposites.

Q3 The function $j(x)$ is also odd. The function $h(x)$ is neither odd nor even.

Q4 a. $t(2) = 2; t(-2) = 2$

b. $t(-1) = -1; t(1) = -1$

c. $t(3) = 7; t(-3) = 7$

d. $t(-0.5) = -1.75; t(0.5) = -1.75$

Q5 $t(a)$ and $t(-a)$ are the same. For example, $t(3) = 7$ and $t(-3) = -7$.

Q6 The functions $u(x)$ and $v(x)$ are also even. The function $w(x)$ is neither even nor odd.

Q7 If it's odd, its behavior on the right side of the origin mirrors its behavior on the left. If it's even, an input to the right of the origin gives the same output as an input the same distance to the left of the origin.

Q8 Graphs of odd functions are symmetrical about the origin. If the graph is rotated by 180° around the origin, it will look exactly the same. (This symmetry is often called *point reflection*.)

Even functions are symmetrical across the y-axis, meaning that reflecting the graph across the y-axis will result in it looking exactly the same.

Q9 $f(x) = 5x, f(x) = x^3 - 2x$, and $f(x) = \sin(x)$ are all odd.

$f(x) = x^2 + 2$ and $f(x) = \cos(x)$ are both even.

$f(x) = x^4 - 3x^3$ is neither even nor odd.

EXPLORE MORE

2.–3. You cannot reflect a graph directly, but steps 2 and 3 provide an easy way to accomplish the same result, by transforming a single point and then constructing the locus of the transformed image. By making the transformed graph thick and light in color, the original dashed graph will remain clearly visible when the function is edited so that the two graphs coincide.

Q10 The results are the same as in Q9 but may be more obvious to students because the transformed image actually appears on the screen.

Q11 An easy way to tell if a polynomial is odd or even is to look at the degrees of its terms (meaning the values of the powers of x). If they're all odd, it's an odd function. This is why odd functions are called "odd." If the degrees are all even, it's an even function. This is why even functions are called "even." Constants (such as the 3 in $f(x) = x^2 + 3$) are considered even-degreed terms since they can be thought of as being a constant times x^0 (for example, $3 = 3x^0$).

There's no such easy way of classifying functions that are not polynomials (such as sin and cos functions).

Q12 If $f(x)$ and $g(x)$ are both odd, $(f + g)(x)$ is also odd. One way to see this would be to redefine $h(x)$ on page 1 as $f(x) + g(x)$ and then explore the resulting dynagraph. Another way is as follows: $(f + g)(-x) = f(-x) + g(-x) = -f(x) + -g(x) = -[f(x) + g(x)] = -(f + g)(x)$. Look at the first and last expressions to see that $(f + g)(x)$ is indeed odd. Using one of these methods, you can see that if $f(x)$ and $g(x)$ are both odd, $(f \cdot g)(x)$ must be even.

The composition $g(f(x))$ is even if f is even, because a and $-a$ both produce the same output from f. Because the output of f is the input to g, the output of g will also be the same for both a and $-a$. If f is odd, the output of f (and the input to g) is opposite for a and $-a$. The result of the composition is odd if g is odd, even if g is even, and neither if g is neither.

If f is neither even nor odd, there is no way to draw any general conclusions about the composition.

WHOLE-CLASS PRESENTATION

Use this presentation to get students to think about characteristics of functions and to view certain types of symmetry that functions can exhibit.

Q1 Open **Odd Even Functions Present.gsp** and ask students to observe the output of the first function (labeled *Even*) as you change the input to its opposite. Press the $a => -a$ button and ask students what they observe. Press the button again to let them watch again while a returns to its original position.

Q2 Ask students to make a conjecture about the behavior of even functions.

Q3 To test the conjecture for many values of a, press the *Show $f(-a)$* button and drag input a. Ask students whether the output behavior supports the conjecture.

Q4 Similarly, use the $b => -b$ button and ask students to make a conjecture about the behavior of odd functions. Test the conjecture for many values of c.

Q5 Use the $c => -c$ button and ask students what pattern they observe in the output. Test for many values of b.

Q6 On page 2, use dragging and/or the buttons to show the behavior of these functions one at a time. Ask students to determine whether each function is odd and to explain their conclusion. Similarly, have students determine which of the functions on page 3 are even.

Q7 On page 4, drag the input marker and ask students to characterize the function $2x - 1$. (It's neither even nor odd.) Edit the function to be $f(x) = 5x$, and ask students to predict whether the function will be even, odd, or neither. Drag the input marker, and then press the *Show $f(-a)$* button and drag again. Ask students to describe in detail what they observed, and ask them to categorize this function as even, odd, or neither. Repeat for the functions below:

a. $f(x) = 5x$ b. $f(x) = x^3 - 2x$

c. $f(x) = \sin(x)$ d. $f(x) = x^2 + 2$

e. $f(x) = x^4 - 3x^3$ f. $f(x) = \cos(x)$

Q8 Ask students what they notice about the four polynomial functions they just categorized. Make a list of which ones were even (d), which were odd (a and b), and which were neither (e). Ask them if they have any ideas about why the names *even* and *odd* are used for the behaviors they observed.

Q9 On page 5, ask students whether they think the function shown is even, odd, or neither. Then ask them which reflection will coincide with the original graph—a reflection across the y-axis or one through the origin. Show each reflection in turn, to check students' answers. Then reset the reflections, edit the function, and repeat.

Finish with a class discussion connecting the behavior of the dynagraphs, the shape of the Cartesian graphs, and the reflections of those graphs.

Inverse Functions

Most arithmetic operations have inverses. If you add 3 to one number to get a second number, you can get back to your original number by adding -3 to your new number. Similarly, if you multiply one number by 2 to get a second number, you can get back to your original number by multiplying by 1/2. We say that -3 is the *additive inverse* of 3 and that 1/2 is the *multiplicative inverse* of 2.

Similarly, many functions have inverses, allowing you to start with the output of a function and get back the input. In this activity you'll explore the inverses of several functions and figure out when inverse functions exist and when they don't.

INVESTIGATE

1. Open **Inverse Functions.gsp.** You'll see two dynagraphs.

Q1 Drag the $f(x)$ input marker to 1.50. What is the output of $f(x)$?

There are no numbers on the axes, but you can observe the value of *x* to move the input marker where you want it.

Q2 Drag the $g(y)$ input marker to the same value as the $f(x)$ output. How does the output of $g(y)$ compare to the input of $f(x)$?

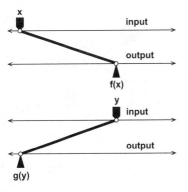

To redefine the function, double-click it and enter a new function definition in the Calculator.

Q3 How can you redefine $g(y)$ so its output is the same as the $f(x)$ input? Do it now.

2. Select the input and output values of $f(x)$ and the input and output values of $g(y)$. Then choose **Graph | Tabulate.** Double-click the table to make the current values permanent.

3. Drag the $f(x)$ input marker to a different value. Then drag the $g(y)$ input marker to match the $f(x)$ output.

Q4 How does the output of $g(y)$ compare to the input of $f(x)$? Write down the input and output of both functions. Double-click the table to record these values.

4. Try three new values of the $f(x)$ input. Each time set the $g(y)$ input to match the $f(x)$ output, and record your values in the table.

Q5 What can you conclude about these two functions?

You can test your conjecture for many input values by attaching the $g(y)$ input to the $f(x)$ output. This is called the *composite function*, written as $g(f(x))$.

5. In this step you'll mark the $f(x)$ output on the $g(y)$ input axis. First drag the $f(x)$ input to a new value. Then select both the $f(x)$ output axis and the $f(x)$ output point (at the top of the output marker), and choose **Construct | Perpendicular Line.** Construct a point at the line's intersection with the $g(y)$ input axis by clicking the **Arrow** tool on the intersection.

6. Attach the $g(y)$ input to this point by selecting the $g(y)$ input point (at the bottom of the input marker) and choosing **Edit | Split Point From Line.** Then select both the intersection and the $g(y)$ input point, and select **Edit | Merge Points.**

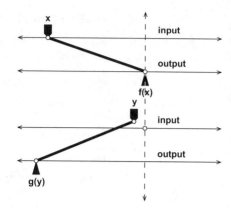

7. Drag the $f(x)$ input to make sure that the attachment worked. Record four new sets of values in your table.

The inverse of $f(x)$ is often written as $f^{-1}(x)$.

If the output of $g(y)$ always matches the input of $f(x)$, f and g are inverse functions.

To change the definition of a function, double-click the function and enter the new definition.

8. Change the definition of $f(x)$ to $x - 1$.

Q6 Are the functions still inverses? If not, how could you change the definition of $g(y)$ to make them inverses again? Change the definition of $g(y)$ to check your answer.

Q7 Change the definition of $f(x)$ to $2x$. What operation do you think you should use so that the output of $g(y)$ matches the input of $f(x)$?

9. Change the definition of $g(y)$, and check your result by trying many values of x.

Q8 What is the inverse of $f(x) = 2x$? Write your answer using f^{-1} notation.

Q9 Change $f(x)$ to $\frac{2}{x}$ and figure out a definition of g that makes f and g inverses. Can you find any values of x for which the input of f doesn't match the output of g? Explain your results.

Q10 Find the inverse of each of the following functions. Check your answers using the dynagraphs, and describe any limitations that you find.

a. $f(x) = 2x - 1$

b. $g(x) = \frac{x}{3} + 1$

c. $h(x) = 3(x + 2)$

d. $p(x) = \frac{(x - 2)}{4}$

e. $q(x) = x^2$

f. $r(x) = x^2 - 2$

Q11 What unusual features do you notice about the last two inverses from Q10?

EXPLORE MORE

Q12 When f and g are inverse functions, what can you say about the behavior of the composite function $g(f(x))$? Use page 2 of the sketch to investigate.

Q13 If g is the inverse of f (written as $g = f^{-1}$), what does that mean about g^{-1} (the inverse of g)? To find out, go to page 3 and follow the directions there.

Inverse Functions

Objective: Students use linked dynagraphs to investigate inverse functions. They define inverses for various functions and test their results by inspecting values in a table and by observing the geometric behavior of the dynagraphs.

Student Audience: Algebra 1/Algebra 2

Prerequisites: None

Sketchpad Level: Easy. Students manipulate a pre-made sketch.

Activity Time: 25–35 minutes

Setting: Paired/Individual Activity (use **Inverse Functions.gsp**) or Whole-Class Presentation (use **Inverse Functions Present.gsp**)

NOTATION

Point out that the variable x in $f(x) = x + 3$ is a placeholder and that $f(y) = y + 3$ represents exactly the same function. Page 1 of the sketch uses $f(x)$ and $g(y)$ to reduce confusion, but the other two pages use $f(x)$ and $g(x)$.

Point out that f is often used as a shorthand for $f(x)$. This shorthand appears in several places in the activity.

Discuss with students the use of $f^{-1}(x)$ to indicate the inverse of $f(x)$. This notation is used in several places in the activity. It's not used throughout because the sketch contains two different function definitions, which may or may not be inverses at any particular time. There's no guarantee that the functions are inverses, so they are identified as f and g rather than as f and f^{-1}.

INVESTIGATE

Q1 When $x = 1.50$, $f(x) = 4.50$.

Q2 When $x = 4.50$, $g(y) = 3.50$.

Q3 To make $g(y) = x$, redefine $g(y) = y - 3$.

Q4 The output of $g(y)$ is equal to the input of $f(x)$. Specific numeric examples will vary.

Q5 The function of $g(y)$ allows you to start with the output of $f(x)$ and get back the input. In other words, $g(f(x)) = x$. Encourage students to refer to their table data in thinking about this question.

Q6 After changing the definition of $f(x)$, the functions are no longer inverses. Changing the definition of $g(y)$ to $x + 1$ makes them inverses again.

Q7 Because x is multiplied by 2, we will have to use division by 2 to form the inverse.

Q8 The inverse of $f(x) = 2x$ is $f^{-1}(x) = x/2$.

Q9 The inverse of $f(x) = 2/x$ is $f^{-1}(x) = 2/x$. In this example, the function is its own inverse. This does not work for the value $x = 0$, because $f(x)$ is undefined when $x = 0$.

Use Q10 and Q11 to introduce the idea that some inverse functions do this successfully for the entire domain of the original function, and others do it for a subset of the domain. If students are familiar with the vertical line test, encourage them to think about how this relates to inverses. (Functions that are invertible over their entire domain also pass a "horizontal line test.")

Q10 Here are the inverse functions:

a. $f(x) = 2x - 1$	$f^{-1}(x) = (x + 1)/2$
b. $g(x) = x/3 + 1$	$g^{-1}(x) = 3(x - 1)$
c. $h(x) = 3(x + 2)$	$h^{-1}(x) = x/3 - 2$
d. $p(x) = (x - 2)/4$	$p^{-1}(x) = 4x + 2$
e. $q(x) = x^2$	$q^{-1}(x) = \sqrt{x}$
f. $r(x) = x^2 - 2$	$r^{-1}(x) = \sqrt{x + 2}$

Q11 The last two functions are inverses only for the domain $x \geq 0$. Outside this domain, the inverses give the wrong result because they return the positive square root. There is no way the inverse function can know when it should use the positive square root and when it should use the negative square root.

EXPLORE MORE

Q12 When $f(x)$ and $g(y)$ are inverse functions, the composite function $g(f(x))$ is the identity function: $g(f(x)) = x$. In other words, it's a function for which the output is always equal to the input.

Q13 If $g(y)$ is the inverse of $f(x)$, then $f(x)$ is also the inverse of $g(y)$. If there's a limitation on the domain of one of the two functions, there's a limitation on the range of the other.

In this presentation you'll use dynagraphs to find and present the inverses of several functions, and to figure out when inverse functions exist and when they don't.

PRESENT

1. Open **Inverse Functions Present.gsp.** Show the directions and buttons. This sketch contains dynagraphs of two functions, $f(x) = x + 3$ and $g(y) = y - 3$.

Q1 Drag the top input marker (x) to change the input variable. Ask students, "For this value of x, what's the output of $f(x)$?"

Q2 Drag the bottom input marker (y) to match the output of $f(x)$. Ask students, "For this value of x, what's the output of $f(x)$?"

> For easier operation, you can press the buttons provided on the sketch rather than dragging the input markers by hand.

2. Select the input and output values of both functions, and choose **Graph | Tabulate.** Double-click the table to make the current values permanent.

3. Drag the x input marker to a different value. Then drag the y input marker to match the $f(x)$ output.

Q3 Ask students how the output of $g(y)$ compares to the input value x. Double-click the table to record these values.

4. Try several new values of x. Each time set y to match the $f(x)$ output and record your values in the table.

Q4 Ask students to make a conjecture concerning these two functions.

You can test your conjecture for many input values by attaching the y input to the $f(x)$ output. This is called the *composite function,* written as $g(f(x))$.

5. Use the *Attach y to f(x)* button, or the directions in steps 5–7 of the student activity sheets, to attach the input of g to the output of f. Drag x again and record four new sets of values.

Q5 Ask students to explain whether the table data confirm their conjecture.

Use page 2 of the sketch to define and examine inverses for several other functions. Pay particular attention to $f(x) = x^2$, and have students describe the problems or limitations they observe in their own words.

Use page 3 to investigate another problem with a surprising feature.

Use page 4 to look at the connection between inverses and the identity function.

Use page 5 to determine the inverse of f^{-1}.

Functions Again and Again

Imagine you have a long piece of licorice you want to split into eight equal pieces to share with seven friends. First, you break the licorice in half. Then, you break the halves in half. And finally, you break the smaller pieces in half for a total of eight equal-sized pieces.

The word for repeating a process over and over again is *iteration*. In the situation described above, you took an object, the licorice, and *iterated* an operation on that object, breaking the piece(s) in half. In this activity you'll iterate arithmetic operations on the coordinates of points.

SKETCH AND INVESTIGATE

To adjust the scale, drag a number on either axis.

1. In a new sketch, choose **Graph | Define Coordinate System.** Adjust the scale until the *x*-axis goes from about −50 to 50 units.

After measuring, the point should be labeled *A*. If not, use the **Text** tool to change its label to *A*.

2. Use the **Point** tool to construct a point anywhere in the plane. Measure the *x*- and *y*-coordinates of the point by choosing **Measure | Abscissa (x)** and **Measure | Ordinate (y).**

Q1 Without doing any calculations or drawings, imagine adding 1 to both the *x*- and *y*-coordinates and plotting the new point. Then imagine adding 1 to both coordinates of the new point to get another new point, and so forth. How would the points be arranged? If you connected them, what shape would they make?

You will use Sketchpad to check your prediction.

Choose **Measure | Calculate** to open the Calculator. Click the measurements in the sketch to enter them into the calculation.

3. Choose **Measure | Calculate** to calculate $x_A + 1$. Similarly, calculate $y_A + 1$. The results are the coordinates of your new point.

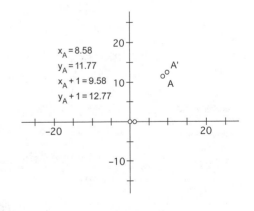

$x_A = 8.58$
$y_A = 11.77$
$x_A + 1 = 9.58$
$y_A + 1 = 12.77$

4. Use the **Arrow** tool to select the new *x*-coordinate $(x_A + 1)$ and the new *y*-coordinate $(y_A + 1)$ in order. Then choose **Graph | Plot As (x, y)** to plot your new point. Label the new point A'.

Select the two measurements and choose **Display | Hide Measurements.**

5. Hide the x_A and y_A coordinate measurements.

Now you will use iteration to create more points.

6. Select point *A* and choose **Transform | Iterate.** You'll get a dialog box that asks you to map point *A* to another object. Click point *A'* in the sketch, to indicate that the same operation should next be applied to *A'*. Click Iterate.

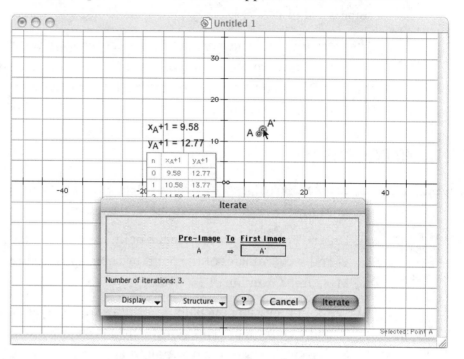

In the sketch itself, Sketchpad plots several more points by repeating (iterating) the operation that maps point *A* to point *A'*. Sketchpad also creates a table showing the calculations for each iterated point.

7. Select one of the iterated points, and press the + key several times. This will increase the number of iterations. Increase it until there are at least ten points.

The sequence is called the *orbit* of *A*.

Q2 The moment of truth has arrived! Drag your initial point *A* around. Describe the sequence of images generated by this iteration.

A function is a set of ordered pairs in which each first element (*x*) is associated with only a single second element (*y*).

Q3 As you drag point *A*, what happens to the values in the table? What changes and what stays the same? Do these values represent a function? Why or why not?

Q4 Imagine a line through your iterated sequence of points. What is the slope of this line?

Q5 Move point *A* as close as you can to (0, 2). What is the *y*-intercept of your imaginary line? Write an equation for the line. Check your result by choosing **Graph | Plot New Function** and plotting your equation. Once you have it right, delete or hide the new function you created.

Now change the iteration rule by changing the way *y*-coordinates are generated. Instead of adding 1 to the *y*-coordinate, what if you were to add 4 or −3?

Functions Again and Again

continued

To change a calculation, double-click it, and then change the expression in the Calculator.

Q6 Change the rule $(y_A + 1)$ to $(y_A + 4)$, and then to $(y_A - 3)$. How do these changes affect both the table and the slope of the imaginary line?

8. You can also change the iteration rule by changing the way you generate new x-coordinates. Try adding 2 to the x-coordinate. Try subtracting 0.5. Can you explain what happens?

Q7 What other combinations of c and d in the iteration rules $(x_A + c)$ and $(y_A + d)$ give the same slope you found in Q4?

Q8 Choose **Graph | Plot New Function.** Plot the equation $f(x) = 3x + 2$. Can you match your iteration to this plot by changing the starting position and the operations that generate new x- and y-coordinates? Describe how you did this.

Q9 Plot the equation $f(x) = -0.5x - 1$. How can you match your iteration to this plot? Delete or hide the function definitions when you are done.

EXPLORE MORE

Q10 How could you change the iteration rules so that the values in the table do not define a function?

So far you've generated only linear sequences of points, but there are many other types of sequences you can generate with this simple iteration rule. Some look quite strange, and some look like functions you're already familiar with.

If you want to increase the number of iterations for this investigation, select the iteration (by clicking on one of its points) and press the + key on your keyboard.

Q11 Instead of adding values to y_A, you could multiply it by a value. For example, instead of using $(y_A + 1)$ you could try $2 \cdot y_A$. (For now, keep adding 1 to the x-coordinate.) You can express this iteration rule more simply like this:

$$x' = x + 1, \quad y' = 2 \cdot y$$

Drag A and describe the pattern of the iterated points. Then investigate the patterns generated by definitions in this form:

$$x' = x + 1, \quad y' = 2 \cdot y + k$$

Q12 Another interesting variation to try is to reverse the coordinates, using a rule like this:

$$x' = y, \quad y' = x + 1$$

Describe this sequence mathematically. Do the coordinates produce a function?

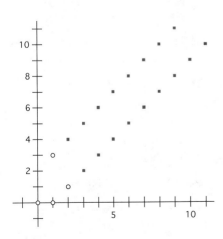

Functions Again and Again

Objective: Students define an iterated coordinate transformation on a point, observe the orbit, decide whether the iterated coordinates represent a function, and describe the function using a linear equation.

Student Audience: Algebra 1/Algebra 2

Prerequisites: Students should be familiar with slope and have a basic understanding of functions.

Sketchpad Level: Intermediate. Students learn how to use Sketchpad's **Iterate** command.

Activity Time: 30–40 minutes

Setting: Paired/Individual Activity (no sketch required) or Whole-Class Presentation (use **Functions Again Present.gsp**)

SKETCH AND INVESTIGATE

In addition to the function that relates the y-coordinate of each point to the x-coordinate, there is another function involved in this activity: the function that transforms a point to its image. This function takes an input point (for instance, point A) and generates an output point (for instance, A'). Such functions can be used to create iterated function systems and can produce striking images like the Barnsley Fern. (See the Barnsley Fern activity in *Exploring Precalculus with The Geometer's Sketchpad*.)

2. After measuring the abscissa, students must deselect the new measurement, then reselect the point and measure its ordinate.

Q1 Answers will vary. It's important for students to try to imagine the results before doing the activity.

6. Students may need to drag the dialog box in order to see the point on which to click.

Q2 The sequence of iterated images is a set of points that lie on a ray starting from point A.

Q3 The values change according to the location of point A. Even though all the numbers change, the difference from one row to the next is always 1. The values do indicate a function, because each x-value corresponds to a single y-value.

Q4 The slope of the line is 1.

Q5 The y-intercept is 2, and the equation is $y = x + 2$.

Q6 When the rule is changed to $(y_A + 4)$, the slope is 4 and the y-values increase by 4 from one row to the next. When the rule changes to $(y_A - 3)$, the slope is -3 and the y-values decrease by 3 from one row to the next. When the rules are $(x_A + 1)$ and $(y_A + k)$, the slope is k.

Q7 The slope in Q4 is 1. The slope of the imaginary line produced by the rules $(x_A + c)$ and $(y_A + d)$ is d/c. Therefore, when $c = d$, the slope will equal 1.

Q8 Answers will vary. Students must change c and d so that $d/c = 3$, and must drag point A to be on the desired line. A logical position for A is at $(0, 2)$.

Q9 Answers will vary. Students must make $d/c = -1/2$ and drag A appropriately—for instance, to $(0, -1)$.

EXPLORE MORE

Q10 If the rule for x is $(x_A + 0)$, all the x-values will be the same and will no longer define a function.

Q11 Repeated multiplication by 2 produces an exponential curve. Starting at $(0, 1)$ produces $1, 2, 4, 8, 16, \ldots$ corresponding to $y = 2^x$. A negative initial y-value results in a graph that curves downward rather than upward. You can generalize this principle using these definitions:

$$x' = x + 1, \quad y' = by, \quad \text{where } b > 0$$

The points fit the curve $y = ab^x$. The y-intercept is a.

The second definition in this section renders a more complicated pattern. Unlike a simple exponential curve, this sequence of points can cross the x-axis.

Q12 The new iteration forms two parallel rows of points. The lines containing the points have a reflection symmetry across the line $y = x + 1/2$. If you construct segment AA' before iterating, you can see the sequence and the symmetry more clearly. The coordinates define a function unless the fractional part of the initial x- and y-values are equal. If they are equal, there will be some x-values that correspond to more than one y-value.

WHOLE-CLASS PRESENTATION

Use the buttons and hints in **Functions Again Present.gsp** to present this activity to the class.

Functions and Relations

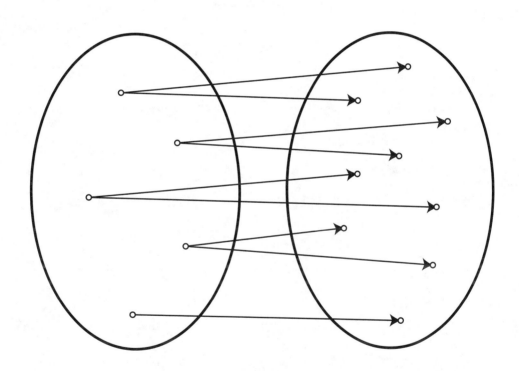

Relations and Functions

A *mathematical relation* exists when two mathematical values are related in some way. A *function* is a specific kind of relation. In this activity you'll explore several relations, decide which of them are functions, and develop a graphical test to determine whether a relation is a function.

EXPLORE

Q1 Think about the mathematical relation between your location and the time. Is it possible for you to be in two different places at the same time? Is it possible for you to be in the same place at two different times? Explain your answers.

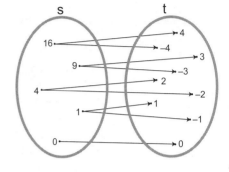

Understanding the answers to these two questions will help you to differentiate between relations that are functions and those that are not.

1. To explore a relation between position and time, open **Relations and Functions.gsp.**

2. Press *Run!* to see Flopsy steal a carrot. Notice the variables *s* (which shows Flopsy's position) and *t* (which shows the time). When Flopsy finishes running, press *Reset.*

To create the table, select measurements *s* and *t* in order, and choose **Graph | Tabulate.**

3. Create a table to record the position and time. Double-click the table to make the first row of values permanent.

4. Press *Dart!* to make Flopsy dart forward. Double-click the table to record her new position and time.

5. Continue moving Flopsy (by pressing *Dart!*) and recording the data (by double-clicking the table) until Flopsy has returned to her original position. The table should contain nine rows of permanent data when Flopsy finishes.

Definition: A *relation* is any set of ordered pairs, such as the ordered pairs (*s*, *t*) that appear in your table.

Q2 Review the data in this relation involving *s* and *t*. Is it possible for Flopsy to be in two different places at the same time? Is it possible for her to be in the same place at two different times? Explain by referring to the data.

6. Page 2 of the sketch shows the same table for *s* and *t*, and also shows two large bubbles containing all the numbers that appear in the table.

Q3 Press the *When was she at s = 4?* button. Answer this question using the arrows that appear. How does this relate to the questions that opened this activity?

7. These two arrows show part of the mapping from position (*s*) to time (*t*). Reset the maps and then use either Show button to show the full mapping.

Q4 From the arrows, are there any ordered pairs that have the same first element (*s*) but different second elements (*t*)? Which ones?

8. Go to page 3. You will look at the same data, in the form (*t, s*) rather than (*s, t*).

Q5 Press the *Where was she at t = 2?* button. Where was Flopsy when the time measurement was 2?

9. The arrows on this page show the mapping from time (*t*) to position (*s*). Reset the maps and then use either Show button to show the full mapping.

Q6 From the arrows, are there any ordered pairs that have the same first element (*t*), but different second elements (*s*)? Which ones?

When a relation is a function, the first element is called the *independent variable*, and the second is the *dependent variable*.

Definition: A *function* is a relation for which there is exactly one second element for each first element. (If *t* is the first element and *s* is the second element, this means that Flopsy cannot be in two different places at the same time.)

Q7 Is the relation (*t, s*) shown on page 3 a function? What about the relation (*s, t*) shown on page 2? How can you tell from the arrows?

EXPLORE MORE

You will now develop a visual test for determining whether a relation is a function.

Q8 Page 4 contains two movable points that are part of a relation. Try to drag the two points so that they both have the same first element but different second elements. Can you do this? Is the relation a function? If not, what values did you use to prove it's not? Where are the points located relative to each other?

Q9 Press the *Problem b* button and try dragging these points. Can you make both of them have the same first element but different second elements? If so, what are the values, and where are the points relative to each other? Is it a function? Also try problems c and d on this page.

Q10 Page 5 shows the graph of a relation, and also shows a movable vertical line. Drag the line back and forth to be sure it remains vertical. How can you use the vertical line to tell whether the graph represents a function?

Q11 Investigate relations b, c, and d on page 5. Which are functions? For each one that's not, at what coordinate did you place the vertical line to prove that it's not?

Relations and Functions

Activity Notes

Objective: Students explore the definitions of relation and function using a model involving position and time, analyze one-to-many and many-to-one mappings, and develop a vertical line test for functions.

Student Audience: Algebra 1/Algebra 2

Prerequisites: It may be helpful (but not critical) if students have already had a brief introduction to relations and functions. The activity includes the definitions of the terms, as well as multiple examples.

Sketchpad Level: Easy. There's very little actual construction to do.

Activity Time: 30–40 minutes

Setting: Paired/Individual Activity (use **Relations and Functions.gsp**) or Whole-Class Presentation (use **Relations Functions Present.gsp**)

EXPLORE

Use this activity when introducing students to the concepts of relations and functions. The example of the rabbit is memorable and will remind students of their conclusion that time cannot be a function of location.

Q1 It is not possible to be in two places at the same time. It is possible to be in the same place at two different times. Explanations will vary, but this can generate an interesting discussion. Teachers generally describe the independent variable as a variable that you have control over—one that you have the ability to vary. But for these measurements, the dependent variable (location) is the one you can control, whereas the independent variable (time) proceeds rudely onward, oblivious to any attempt to control it.

Q2 Flopsy cannot be in two locations at the same time, but she is at position 9 (for instance) at both $t = 1$ and $t = 7$. Thus she's at the same place at two different times.

Q3 Both arrows lead away from 4 in the position bubble, leading to 2 and to 6 in the time bubble. Thus Flopsy was at this position at two different times: two seconds after starting and six seconds after starting.

Q4 Several values of position correspond to more than one ordered pair (in other words, to more than one

row of the table). There are two pairs with a first element of 16, two with a first element of 9, two with a first element of 4, and two with a first element of 1.

Q5 When $t = 2$, Flopsy was at position 4.

Q6 There are no ordered pairs with the same time and different locations—that is, none with the same first element and a different second element. This is the fundamental definition of a function.

Q7 The relation (t, s) on page 3 is a function. The relation (s, t) on page 2 is not. You can tell from the arrows because there's only a single arrow leading from any given first element.

EXPLORE MORE

Q8 You can drag the points so that they have the same first element and different second elements. For instance, the ordered pairs (3.50, 1.55) and (3.50, 6.72) are both part of the relation. Therefore, this relation is not a function. These points are aligned with each other vertically.

Q9 Problem b is a function; there are no ordered pairs with the same first element and different second elements. Problem c is also a function. Problem d is not a function; numeric data to demonstrate this will vary, but students should observe that they are able to arrange the two points so they are aligned vertically.

Q10 This is not a function, because there are several places where the same first element corresponds to more than one second element. The vertical line demonstrates this condition by intersecting the function plot in two or more places.

Q11 Problem b is a function, and the vertical line can never intersect it at more than a single place. Problem c is the inverse of the cosine function; this inverse is not a function itself, because the vertical line intersects it in many places. The inverse cosine must be defined to have a restricted range if it is to be dealt with as a function. Problem d is a step function. Though the vertical line comes close to two different segments of the graph at certain positions, in fact it never intersects the two segments at the same time.

Exploring Algebra 2 with The Geometer's Sketchpad
© 2007 Key Curriculum Press

2: Functions and Relations 35

The presentation should follow the sequence in the student activity. Begin with an informal description of relations and functions, and use the presentation to motivate a more precise definition.

Start with these descriptions: A *mathematical relation* exists when two mathematical values are related in some way. A *function* is a specific kind of relation.

EXPLORE

Q1 Can you be in two different places at the same time? Can you be in the same place at two different times? Ask a number of students to respond and explain.

1. Open **Relations and Functions Present.gsp** and press *Run!* to see Flopsy steal a carrot. Run the animation again and point out the variables *s* (which shows Flopsy's position) and *t* (which shows the time).

2. Reset Flopsy. To record her position and time, press *Show Table* and double-click the table to make the first row of values permanent.

3. Press *Dart!* repeatedly to make Flopsy dart forward. At the end of each movement, double-click the table to record the new position and time. You should have nine rows of permanent data when you finish.

> This table does not define a function. Our purpose is to clarify the distinction, so the first relation students work with should not be a function.

Give students a more precise definition, now that it's illustrated by a table: A *relation* is any set of ordered pairs, such as the ordered pairs (*s*, *t*) that appear in your table.

6. On page 2, press the *When was she at s = 4?* button. Question students, and discuss the meaning of the arrows, until they are all convinced that there's not just one right answer to this question.

7. These two arrows show part of the mapping from position (*s*) to time (*t*). Reset the maps and then use either Show button to show the full mapping.

> For a relation to be a *function*, each input value must correspond to only a single output value.

8. Contrast the full mappings (with all arrows showing) from pages 2 and 3 to further clarify the distinction between the relation that is a function (page 3) and the one that is not (page 2).

9. Use page 4 to see which problems allow ordered pairs that violate the function definition and to point out that the two ordered pairs are aligned vertically when they prove a relation is not a function.

10. Use page 5 to demonstrate the vertical line test for functions.

Finish with a class discussion reviewing the definitions. Ask students why they think functions might be so important in mathematics. (One benefit of working with functions is that the ability to find a unique output, given any input, makes many problems much easier to manage.)

The Circumference Function

In a *function,* one quantity depends on another quantity, just as the number of songs a jukebox plays depends on how much money is put in. In the case of a circle, we can say that its circumference *depends* on its diameter—the farther it is across a circle, the farther it is around it. In this activity you'll explore this connection as a *functional* relationship between two changing quantities.

IMAGINE AND PREDICT

Imagine using a compass to draw a circle. You measure the circle's diameter (1.36 cm) and calculate its circumference (4.27 cm). You then plot the point (1.36, 4.27) on a piece of graph paper. You draw a second circle, measure its diameter (*d*), calculate its circumference (*c*), and plot a second point (*d, c*). You do this for many more circles.

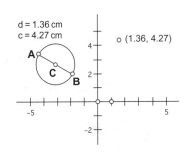

Q1 What will your graph look like after you plot many points? Discuss this with your group. Will the graph be straight or curved? Why? If it's straight, what will its slope be? If it's curved, will it bend up or down? Will it go forever in both directions, or will it start or stop somewhere? Write down your predictions.

SKETCH AND INVESTIGATE

The **Segment** tool

The **Compass** tool

Now test your prediction using Sketchpad.

1. In a new sketch, construct a segment using the **Segment** tool. With the segment selected, choose **Construct | Midpoint.**

2. Construct a circle whose diameter is the segment. To do this, choose the **Compass** tool and click first on the midpoint and then on one endpoint. Drag each endpoint of the diameter to make sure your construction holds together.

3. Measure the segment's length by selecting it and choosing **Measure | Length.** Similarly, select the circle and measure its circumference.

4. Label the diameter measurement *d* and the circumference measurement *c*.

Double-click a measurement with the **Text** tool (the A in the Toolbox) to edit its label.

5. Plot the point (*d, c*) by selecting the measurements in order and choosing **Graph | Plot As (x, y).** The plotted point appears.

You can drag the tick mark numbers on the axes to change the scale of the graph.

6. Select the plotted point and choose **Display | Trace Plotted Point.**

Q2 The moment of truth has arrived. Drag either endpoint and observe the point being traced. Describe this trace on your paper. Was your prediction correct?

The Circumference Function

continued

FURTHER INVESTIGATION

You've now seen the shape of the diameter-circumference trace. But why does it look the way it does? You may know that lines can be modeled with equations of the form $y = mx + b$, or $c = md + b$ in this case. So what are m and b for this line?

Q3 Why does it make sense that the trace goes through the origin? What does this tell you about m and b in the equation $c = md + b$?

Q4 Drag the diameter endpoints so that d is as close to 1.00 as you can make it. What is the approximate circumference of the circle? Is this number familiar to you? If so, what's it called?

Choose
Measure | Calculate
to open the Calculator.
Click on c and d in the
sketch to enter them
into the calculation.

Q5 Use Sketchpad's Calculator to find the ratio c/d. Drag one of the diameter endpoints and observe the ratio. What happens and why?

7. Turn off tracing for your plotted point.

8. Construct a ray from the origin through the plotted point. Then measure the ray's slope.

Q6 What does the slope measurement tell you about m and b in $c = md + b$?

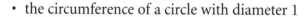

$d = 1.52$ cm
$c = 4.77$ cm

You've now seen the same number in three places:

- the circumference of a circle with diameter 1

Choose **Graph | Plot New Function** and enter your proposed expression. Use x for the diameter (because you used d for the x-coordinate earlier).

- the ratio of any circle's circumference to its diameter

- the slope of the graph of a circle's circumference as a function of its diameter

Q7 Use your results for m and b from Q3 and Q6 to plot a line that includes the ray from step 8. On your paper, write the function you used.

EXPLORE MORE

Q8 What's an appropriate domain for the circumference function? Select the function plot from Q7 and choose **Properties** from the Edit menu. Go to the Plot panel and set an appropriate domain for this situation.

Q9 Consider a circle's *area* as a function of its radius. How will the plot of that function compare with the diameter-circumference plot? Make a prediction, then use the techniques from this activity to confirm.

The Circumference Function

Objective: Students construct a circle based on its diameter. They then measure, graph, and analyze the function that connects the diameter and circumference of a circle.

Student Audience: Pre-algebra/Algebra 1/Algebra 2

Prerequisites: Students should have had a basic introduction to functions.

Sketchpad Level: Intermediate

Activity Time: 30–40 minutes. The time will vary depending on students' Sketchpad experience, since they do all the construction steps themselves.

Setting: Paired/Individual Activity (no sketch needed) or Whole-Class Presentation (use **Circumference Fn Present.gsp**). The pages of **Circumference Fn.gsp** show the activity as it's supposed to be after steps 4, 6, Q4, Q5, and Q7.

Related Activities: Relations and Functions (to introduce the concepts), Radius and Arc Length (measuring and analyzing an existing construction), Functions in a Triangle (also involving construction), Functional Geometry (finding functions in a variety of existing constructions)

IMAGINE AND PREDICT

One of the interesting things about this activity is the relationship between *formulas* and *functions*. Most students at this level know the formula $c = \pi \cdot d$, and most can graph the line $y = 3x$ or $y = 3.14x$ or even $y = \pi x$. Yet very few are able to recognize the first formula as being equivalent to the final equation. In this activity, students are challenged to think of c as a dependent variable just like y, and of d as an independent variable just like x. Rich connections can result from a lively classroom discussion.

Q1 It is important that students take the opening part of the activity—Imagine and Predict—seriously. This gives them an investment in the activity and exercises an important part of their mathematical thinking. The key issue here isn't right or wrong answers, it's the quality of the explanations. Use your judgment about providing hints during this section; too much information may diminish some students' interest in the rest of the activity.

SKETCH AND INVESTIGATE

2. Students need to click first on the midpoint and then on one of the endpoints for the circle to be attached properly. The most common mistake students make is to click the second time in blank space in such a way that the circle may *appear* to pass through the endpoint but is not *constructed* to do so. Dragging either endpoint will reveal this problem.

3. The measurement commands won't be available if too many objects are selected. Students can deselect all objects first by clicking in blank space.

Q2 The trace is a ray, emanating from the origin. More details to come in subsequent answers!

FURTHER INVESTIGATION

Q3 It makes sense that the graph goes through the origin, because a circle with a diameter of zero would have a circumference of zero. If students don't accept that a zero diameter, zero circumference circle exists, use a limit argument: Drag an endpoint so the circle has a tiny diameter, and watch the two measurements approach zero as the plotted point approaches the origin. If the two endpoints actually coincide, the construction and measurements disappear. (This is a nice situation in which to investigate a limit and see why it's important to make the quantity as small as you can—but *not* to make it zero.)

That the trace goes through (or at least to) the origin tells us that the y-intercept (b) is 0. However, this fact tells us nothing about the slope (m).

Q4 The circumference is approximately 3.14. This number is π (to the nearest 0.01). Students can use the properties of the measurement (**Edit | Properties | Value**) to increase the precision, but the accuracy of the result still depends on how close to 1 they can make the diameter. Consider challenging students to find a method of constructing a circle with a diameter of *exactly* 1.

Q5 The ratio stays the same—a close approximation of π—no matter the size of the circle. This approximation is closer than the one from the previous question, because the diameter is incorporated into the calculation. The result doesn't depend on dragging the diameter to a precise value.

8. Students can use the **Ray** tool to construct the ray. (Press on the current **Straightedge** tool, then drag and release over the **Ray** tool in the palette of tools that appears.) Or they can use **Construct | Ray** after selecting the origin and the plotted point.

Q6 It tells us that $m = \pi$ since m represents a line's slope. The construction assumed that $b = 0$, because the ray started from the origin.

Q7 The function students should plot is $f(x) = \pi \cdot x$.

EXPLORE MORE

Q8 The appropriate domain is either $x > 0$ or $x \geq 0$. (This distinction could make for an interesting discussion!)

Q9 The plot will be a parabola. Specifically, it will be the parabola with the equation $y = \pi x^2$, which corresponds with the familiar area formula $A = \pi r^2$. It has the same domain as the linear graph.

WHOLE-CLASS PRESENTATION

Use **Circumference Fn Present.gsp** to present this activity to the whole class. Use the animations provided in that sketch to view the behavior and measurements, answer questions, and stimulate a class discussion.

Radius and Arc Length

In this activity you'll explore the relationship between the radius and the arc length of a semicircle, and decide whether this relationship is a function.

Suppose you want to build a moving sculpture for your sister's birthday. The sculpture will be made of eight wheels of different sizes, with a string attached to each wheel. When the sculpture is in motion, a balloon attached to the other end of each string flies upward, pulling the string and spinning the wheel one-half of a revolution. To properly design the sculpture, you need to know how the height each balloon will reach is related to the radius of its wheel.

RADIUS-ARC LENGTH RELATIONSHIP

1. Open **Radius and Arc Length.gsp.** Press *Go Up* to watch a single wheel and balloon in action.

2. The actual sculpture will have eight wheels of different radii. Set a different radius by pressing *Go Down* and then dragging point *C*.

3. Operate this different-size wheel by pressing *Go Up*.

Q1 What do you observe?

4. Repeat steps 2 and 3 six more times. Try to space the wheel sizes fairly evenly.

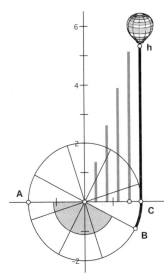

Four traces are finished and a fifth trace is in progress.

You should have eight traces left showing where the balloon went.

Q2 Describe how the height reached by the balloon changes as the radius of the circle changes.

Q3 If you draw a line through the tips of the traces, will it be straight or curved?

Next, you will collect some data on the circle radius and string length.

5. Erase the traces by choosing **Display | Erase Traces.**

6. With the balloon in its upper position, measure the *x*- and *y*-values of point *h*.

7. Put these measurements in a table by selecting them and choosing **Graph | Tabulate.** Enter the current values permanently by double-clicking the table.

8. Operate the balloon again with a different-size circle. When the motion stops, record the next set of coordinates by double-clicking the table.

You can erase the traces at any time by choosing **Display | Erase Traces.**

Select the point and choose **Measure | Abscissa (x).** *Select the point again and choose* **Measure | Ordinate (y).**

9. Repeat step 7 until you have data from eight circles of different sizes.

10. Plot the data from the table by selecting it and choosing **Graph | Plot Table Data.** Make sure the options are correct before closing the dialog box.

Q4 How are the plotted points arranged? Do they appear to be in a straight line? What else do you observe about their arrangement?

Use the **Line** tool, or select the two points and choose **Construct | Line.**

11. Construct a line through the two plotted points that are farthest apart.

Q5 Does this line appear to go through all the other points? Check your answer by measuring the distance from several of the other points to the line. (To do this, select one point and the line, and choose **Measure | Distance.**) What do your results tell you?

Q6 Measure the equation of the line. What are the slope and intercept?

Q7 What does the value of the intercept tell you? Explain why this makes sense.

Q8 Where have you seen the value of the slope before? Explain why this number makes sense.

Q9 Is the relation between the radius and the string length a mathematical function? Why or why not?

Q10 Write the equation for the function using l for string length and r for radius.

Q11 Write equations representing the functions (r, c) and (d, c), where c is the circumference and d is the diameter. Explain how you can figure out each of these equations from the equation you wrote for Q10.

EXPLORE MORE

12. On page 2, operate the wheel by hand by dragging B. Notice the measurements that show the angle and distance through which you have moved B.

After plotting the data, adjust the scales on the x-axis and y-axis to show the data better.

Q12 Do you think there's a simple relationship between the angle and the length of the arc? Leave the radius constant, gather data in a table, and plot the table data. What do you observe about the plotted data? What is the equation of the data? Can you find an interpretation for the constant in this equation?

Your result will reveal an angle that's important in later mathematics.

Q13 Drag B so that the arc length is as close to the radius as you can make it. What's the arc angle? Tabulate the radius, arc length, and arc angle for several different values of the radius, each time making the radius and arc length as close to equal as you can. What do you observe about the arc angle?

Radius and Arc Length

Objective: Students use a Sketchpad model to explore the relationship between the radius of a circle and the arc length of a semicircle to decide whether it is a mathematical function, and if so to find a mathematical equation for it.

Student Audience: Algebra 1/Algebra 2

Prerequisites: None

Sketchpad Level: Easy to Intermediate. Students manipulate a pre-made sketch and tabulate and plot data.

Activity Time: 30-40 minutes

Setting: Paired/Individual Activity (use **Radius and Arc Length.gsp**) or Whole-Class Presentation (use **Radius and Arc Length Present.gsp**)

RADIUS-ARC LENGTH RELATIONSHIP

Q1 The second wheel made a different vertical bar, in a different horizontal position and of a different height.

Q2 The farther point C is dragged from the origin, the higher the traced bar is. In other words, the greater the radius of the wheel, the longer the string is and the higher the balloon goes.

Q3 The line appears to be straight.

Q4 The plotted points appear to be in a straight line, and it appears that the line may also go through the origin.

Q5 The constructed line does appear to go through all the points. The distance measurements will confirm this; they should all be zero.

Q6 The equation is approximately $y = 3.14159x + 0.00000$. The slope is 3.14159, and the intercept is very nearly zero. (These results will be more accurate if students use points far apart, and will be less accurate, perhaps significantly so, if they use points that are very close to each other.)

Q7 The intercept of zero indicates that the line goes through the origin. This makes sense: If the radius of the wheel were zero, the length of string would also be zero. Because the line goes through the origin, this function is an example of *direct variation*.

Q8 The value of the slope is π. This makes sense because the circumference of a circle is $2\pi r$, so the arc length of half a circle should be πr.

Q9 This relation is a mathematical function, because each possible value of the radius is associated with only a single string length.

Q10 The equation is $l = \pi r$. Using function notation, students could also write $l(r) = \pi r$.

Q11 The equations are $c = 2\pi r$ or $c(r) = 2\pi r$ and $c = \pi d$ or $c(d) = \pi d$. To figure the first equation out from the equation in Q10, multiply by 2 because the entire circumference is twice as long as the string. To figure the second equation out from the equation in Q10, you must realize that the diameter is also twice as long as the radius. Because both numbers are twice as big, the same equation works.

EXPLORE MORE

Q12 There is a simple linear relationship between the angle and the length of the arc. Like the previous example, this is direct variation. The equation is *length* $= k \cdot angle$, where k is a constant that depends on the radius. In fact, $k = \pi r/180$ because at 180° the arc will be a semicircle and the arc length will be equal to πr.

Q13 No matter the size of the wheel, the arc angle must be close to 57.3° to make the arc length equal to the radius. This angle is called a *radian* and is very important in later mathematics.

WHOLE-CLASS PRESENTATION

Use the Presenter Notes and **Radius and Arc Length Present.gsp** to present this activity to the whole class.

Radius and Arc Length

In this presentation you'll gather and plot data about the radius and arc length of a semicircle, decide whether the relation is a function, and measure the equation.

Start by describing the model, as in the introduction to the student activity pages.

RADIUS-ARC LENGTH RELATIONSHIP

1. Open **Radius and Arc Length Present.gsp.** Press *Go Up* to show the wheel and balloon in action.

To change the radius, drag point C or press Increase Radius or Decrease Radius.

2. The actual sculpture will have eight wheels of different radii. Set a different radius by pressing *Go Down* and then increasing or decreasing the radius.

3. Operate this different-size wheel by pressing *Go Up.*

Q1 Ask students what they observe about the height.

4. Measure the height, and mark it by pressing the *Mark Height 1* button.

Double-click the table to record the current values permanently.

5. Make a table to record the data by selecting the radius and height measurements and choosing **Graph | Tabulate.**

6. Make at least four more measurements, each time recording the data in the table and marking the height of the balloon with one of the numbered points.

*You can erase the traces at any time by choosing **Display | Erase Traces.***

You should have at least five traces left showing where the balloon went. You should also have at least five rows of data in your table and five of the height traces marked by numbered points.

Q2 Ask students to use the table to decide whether the relation is a function.

Q3 Ask whether a line through the tips of the traces will be straight or curved.

*Use the **Line** tool or the **Construct | Line** command.*

7. Construct a line through two of the height markers.

Q4 Ask students whether the line appears to go through all the data points or only some of them. Ask them to estimate the slope of the line and the *y*-intercept.

*Select the measurement, choose **Edit | Properties,** and change the precision in the Value panel.*

Q5 Measure the line's slope. Change the precision of the measurement to hundred-thousandths. Ask students whether they've seen this number before.

Q6 Ask students to write the equation of the line. Then measure the equation and compare their equations with the measured one.

Use page 2 to investigate the angle of rotation for which the arc length is equal to the radius. Follow the directions on the sketch.

You may also want to mention that students will see this angle later.

Q7 Ask students how this angle depends on the radius. What do they think about their results?

Q8 Tell students this angle is called a *radian;* ask how they think it got its name.

Functions in a Triangle

In this activity you'll construct a triangle, make two measurements, change the shape of the figure, and observe how these measurements are related. This kind of relationship—where one quantity changes as a result of another quantity changing—is an example of a function.

CONSTRUCT A TRIANGLE

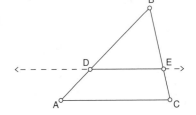

Choose **Edit | Preferences,** click on the text tab, and check "For All New Points."

1. In a new sketch, change Sketchpad's preferences to label points automatically as you construct them.

2. Construct △*ABC* using the **Segment** tool.

3. Construct point *D* on side *AB* by using the **Point** tool.

Select *D* and $\overline{AC}$. Then choose **Construct | Parallel Line.**

4. Construct a line through point *D* parallel to $\overline{AC}$.

5. Construct point *E* where this line intersects $\overline{BC}$. To do this, click on the spot with the **Arrow** tool.

6. Hide the parallel line and construct $\overline{DE}$.

7. Measure the distance from *B* to $\overline{DE}$ by selecting the point and segment and choosing **Measure | Distance.** Then measure the length of $\overline{DE}$ by selecting the segment and choosing **Measure | Length.**

FUNCTIONAL RELATIONSHIPS

Now that you've measured your construction, explore the relationship between the two measurements.

8. Drag point *D* back and forth. Observe how the measurements change and think about how they relate to each other.

Q1 Is it possible to drag any of the points in a way that makes one measurement get larger while the other one gets smaller? What can you conclude from dragging?

The behavior of these measurements suggests that they relate to each other according to a rule. To investigate this rule and decide whether it's a mathematical function, you can graph it.

First, you must decide which of the two measurements to use as the independent variable. You can (a) make the distance from *B* to $\overline{DE}$ the independent variable and the length of $\overline{DE}$ the dependent variable, or (b) make $\overline{DE}$ the independent variable and *B* to $\overline{DE}$ the dependent variable.

Q2 Decide which of these choices (a or b) makes the most sense to you. Explain your reasoning.

Q3 On your paper, draw an x- and a y-axis, and a prediction of what you think a graph of the relationship might look like. Don't bother to scale the axes or to plot points of actual measurement values. Use only your intuitive sense of how the measurements change in drawing your prediction.

If you don't see the plotted point right away, make your triangle smaller or drag any one of the tick mark numbers on the x-axis toward the origin.

9. Select the two measurements—first the independent variable, then the dependent variable—and choose **Graph | Plot As (x, y).** A set of axes and a plotted point will appear. Drag point D and observe the path of the plotted point on the grid.

10. To see the set of all possible plotted points, select the plotted point and choose **Display | Trace Plotted Point.** Then drag D again, moving it all the way from A to B to see the entire graph. (You can erase the traces at any time by choosing **Display | Erase Traces.**)

Q4 How close was your prediction? What does the graph's shape tell you about the functional relationship in the triangle?

The *domain* of a function is the set of possible values for the independent variable. The *range* is the set of possible values for the dependent variable.

Q5 Describe the domain and range of this function. What do the maximum values of each correspond to in the construction?

11. To make the graph permanent, select both D and the plotted point, and choose **Construct | Locus.** Also turn off tracing for point D.

12. Drag any vertex of the triangle to change the triangle's shape and size. Observe how this affects the graph.

Q6 Describe the different types of graphs you get with different kinds of triangles. For example, what shapes give graphs with steeper slopes? Less steep slopes? Try to come up with general rules that describe how the function changes as you alter the shape of the triangle.

EXPLORE MORE

With the New Function dialog box open, click on the measurements in the sketch to enter them into the function equation.

13. Choose **Graph | Plot New Function.** Make use of the measurements in the sketch to plot a function that coincides with your locus graph.

Q7 How do you think the plots of length $\overline{BD}$ versus length $\overline{DE}$ or length $\overline{BE}$ versus length $\overline{DE}$ would compare to your existing plot? Draw a graph of your prediction on paper. Then test it by measuring the distances and plotting a point as you did before.

Q8 Make other measurements in the construction and investigate other functional relationships. See if you can find one that's not linear.

Functions in a Triangle

Objective: Students measure constructions in a triangle, investigate the relations that result, and plot graphs to study how they behave and whether they are functions.

Student Audience: Pre-algebra/Algebra 1/Algebra 2

Prerequisites: Some basic familiarity with graphing in the *xy* plane

Sketchpad Level: Intermediate. This is a good first activity for students learning to make their own sketches.

Activity Time: 25–35 minutes

Setting: Paired/Individual Activity (no sketch needed). The pages of **Triangle Functions.gsp** show the activity at steps 5, 7, and 10.

Related Activities: The Circumference Function is similar to this activity; it also involves creating a construction from scratch and investigating relationships between various measurements. By constructing the geometric figure from scratch, students are involved more intimately in the geometric behavior and functional relationships they explore. They are more likely to remember the insights they gain about what a function is and how functions work because they have created the basis for the activity themselves.

The activity Relations and Functions is a good introduction to the difference between relations that are functions and relations that are not. The Functional Geometry activity asks students to make and analyze measurements on a variety of pre-made sketches, with a lot of emphasis on the choice of independent and dependent variables.

CONSTRUCT A TRIANGLE

The key geometric relationship in this activity is that △DBE is similar to △ABC. As D travels along segment AB, △DBE forms an infinite number of similar triangles. Since corresponding parts of similar triangles are proportional, the ratio of the length of $\overline{DE}$ to the distance from B to $\overline{DE}$ will always be the same. This ratio will be the slope in the linear function.

Students who have taken high school geometry should understand and be able to discuss this geometric relationship. Students without this background should also understand it, albeit less formally.

2. Students need to be sure to click on existing endpoints of segments when drawing the second and third sides of the triangle. At the end of this step, they should have three segments and three points (labeled *A*, *B*, and *C*) in their sketches. (If students make and then delete mistakes, their labels may no longer match the descriptions here. In this case they should use the **Text** tool to change the labels to match the directions.)

3. Make sure $\overline{AB}$ is highlighted before you click. Drag the new point with the **Arrow** tool after this step to confirm that it moves only along the segment.

5. An alternate way to construct the point of intersection is to select the line and the segment and choose **Construct | Intersection.**

6. To hide the line, select it and choose **Display | Hide Line.** Use the **Segment** tool to construct $\overline{DE}$, or select *D* and *E* and choose **Construct | Segment.**

FUNCTIONAL RELATIONSHIPS

Q1 No matter what you drag, it's not possible to make one measurement increase while the other decreases. This suggests that there's a mathematical rule governing the behavior of the measurements.

Q2 Either choice is acceptable, and class discussion will be improved if different students make different choices here. Most students will probably choose *B* to $\overline{DE}$ as the independent variable because that measurement seems to change more directly as point *D* moves up and down $\overline{AB}$.

Q3 Answers will vary. Encourage students to put thought into their prediction. Ask them to explain their choices in writing or to compare and discuss them in groups. The graphs of students who chose different independent variables will be especially interesting throughout the activity.

Q4 The graph is a straight line—or, more specifically, a straight line segment—with one endpoint at the origin and the other somewhere in the first quadrant. The fact that the graph is linear tells you that as one quantity changes at a consistent rate, so does the other, though one may change faster than the other. The fact that it goes through (or at least up to) the origin means that when the distance between B and $\overline{DE}$ is zero, so is the length of $\overline{DE}$, and vice versa.

Q5 For choice a (the distance from B to $\overline{DE}$ is the independent variable), the domain is the set of distances from zero to the height of $\triangle ABC$, inclusive. The minimum distance, zero, occurs when D is at B. The maximum value corresponds to the height of $\triangle ABC$ and occurs when D is at A and $\overline{DE}$ coincides with $\overline{AC}$. For choice b, the above is true of the range.

For choice a, the range is the set of lengths from zero to the length of $\overline{AC}$, inclusive. The minimum value, zero, occurs when D is at B. The maximum value is equal to the length of $\overline{AC}$ and occurs when D is at A and $\overline{DE}$ coincides with $\overline{AC}$. For choice b, the above is true of the domain.

Q6 For choice a, the wider the base is relative to the height, and the shorter and wider the triangle is, the steeper the slope will be. (For choice b, the slope will be less steep.) This is true because the length of $\overline{DE}$ changes very quickly for very small changes in the distance between B and $\overline{DE}$. The smaller the base is relative to the height, and the taller and skinnier the triangle is, the less steep the slope will be for choice a. (It will be more steep for choice b.) This is true because as the distance from B to $\overline{DE}$ changes a lot, the length of $\overline{DE}$ changes much more slowly. Since the locus of choice b behaves like a reflection of choice a, the opposite of the preceding will be true for choice b.

Furthermore, any pair of similar triangles would produce graphs with the same slope. The only difference is that the larger triangle would have a larger domain and range.

EXPLORE MORE

13. Since the slope of the segment is the ratio of the two measurements, the function to plot is

$f(x) = (m\overline{DE}/\text{Distance } B \text{ to } \overline{DE}) \cdot x$ for choice a, and

$f(x) = (\text{Distance } B \text{ to } \overline{DE}/m\overline{DE}) \cdot x$ for choice b.

To make the plot coincide exactly with the trace of the plotted point, change the function's domain using **Edit | Properties | Plot.**

Q7 These plots will be similar to the original plot—segments from the origin—but with slopes less than or equal to that of the original plot. This is because the shortest distance from B to $\overline{DE}$ is along a perpendicular, and this is how the distance from B to $\overline{DE}$ is measured. Since the ratio corresponding to the slope of the plots has BD, BE, or the distance from B to $\overline{DE}$ in the denominator, the shortest distance corresponds to the greatest slope. The slopes would be equal for right triangles. (For example, if C were a right angle, the length of $\overline{BE}$ would equal the distance from B to $\overline{DE}$, and these two plots would coincide.)

Q8 Relationships between linear values (height, for example) and the area will be quadratic.

Functional Geometry

In many geometric figures, different aspects or quantities change in related ways. When two things change in a related way, their connection is called a *relation*. For example, the diameter and radius of a circle change together, so they form a relation.

Sometimes the second thing in a relation not only changes with the first, but actually depends on it. In this case we say the relation is also a *function*. In this activity you'll explore relations between geometric measurements and determine which of those relations are functions.

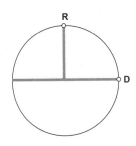

RADIUS AND DIAMETER

Q1 Does the diameter of a circle depend on the radius? Is the relation between the diameter and the radius a function? Explain your reasoning.

Q2 Does it also work the other way? Can you also think of the radius as a value that depends on the diameter? Is this relation a function? Explain.

1. Open **Functional Geometry.gsp.** Familiarize yourself with the figure by dragging each of the two points.

To measure these aspects of the circle, select each one at a time and choose **Measure | Length.**

2. Measure the circle's radius and diameter, and decide which measurement you want to control directly. This is the independent variable, and the other measurement is the dependent variable.

Q3 Which measurement will you use as the independent variable? Which point must you drag to change it?

Q4 As you drag your chosen point, observe how the two values change together, and try to imagine what a graph of the function would look like. Draw your prediction on your paper.

If the plotted point isn't visible, drag one of the tick numbers closer to the origin.

3. To test your guess, select the two measurements in order (independent variable first, then dependent variable) and choose **Graph | Plot As (x, y).** Drag your chosen point and observe the behavior of the plotted point.

4. Select the plotted point and choose **Display | Trace Plotted Point.** Observe the path of the plotted point as you drag your chosen point.

Q5 Compare your guess with the actual graph. Why is the graph the shape it is?

Q6 Think back to your decision to designate one of the measurements as the independent variable. Did it matter which one you chose? Why or why not? How are the graphs different? To answer this question, repeat steps 2–4, but this time make the opposite choice.

Q7 Based on what you've learned, is the relation a function? Why or why not?

INVESTIGATE

This sketch has eight more geometric figures on the remaining pages. On each page, explore how two different aspects of the figure relate to one another.

Measure two aspects of the figure that you're interested in exploring, and use these measurements to complete steps 2–4 and to answer Q3 and Q4.

Compare your prediction with Sketchpad's graph of the function. Why did the graph take the form that it did? Did it matter which measurement you designated as the independent variable? Why or why not?

Repeat the process, but this time make the opposite choice, and compare the two graphs.

Knowing what you've learned about the two aspects you've chosen in the figure, is either of the relations a function? Explain your answer.

SOME FUNCTION SUGGESTIONS

Page	Suggested Measurements
Circle	diameter, radius, circumference, area
Triangle	AD, DE, distance from A to DE
Xquad	AD, DE
Rect 1	AD, AB, perimeter, area
Rect 2	AD, AB, perimeter, area
Polygons	AD, area, perimeter for each polygon (Compare the graphs.)
Ladder	AD, AB
Hanger	AD, $\angle BDC$, area, perimeter
Tube View	tube length, tube diameter, distance, field of view

EXPLORE MORE

5. See if you can write an equation for one or more of the functions you graphed. Plot that function (choose **Graph | Plot New Function**) to see if it matches the locus graph (choose **Construct | Locus**). You might need to measure things in the sketch to use as parameters in your function.

© 2007 Key Curriculum Press

Functional Geometry

Objective: Students explore relations defined by geometric measurements. They create graphs of the relations, explain how they decided which would be the independent and which the dependent variable, and determine whether the relation is a function.

Student Audience: Algebra 1/Algebra 2

Prerequisites: Students should be familiar with graphing in the *xy* plane and with functions. Several other activities (including The Circumference Function and Functions in a Triangle) can be used to introduce these topics.

Sketchpad Level: Intermediate. Students measure objects, plot the resulting measurements, and create a graph.

Activity Time: 25–50 minutes. The time required depends on how much exploration students do. Try to allow an entire class period, and encourage students to explore and discuss their own functions and to plot equations to match their findings.

Setting: Paired/Individual Activity (use **Functional Geometry.gsp**)

Use this activity to encourage students to explore examples of relations and functions on their own. They will find many interesting functional relationships. There are even some quartic (4th-degree) polynomials lurking for those curious enough to find them.

INDEPENDENT AND DEPENDENT VARIABLES

This activity provides an excellent opportunity to explore and discuss the terms *independent variable* and *dependent variable*, so make sure that students plot each pair of quantities both ways, first with one quantity as the independent variable, and then with the other.

In the class discussion, point out that the terms have both a strict mathematical meaning and a common meaning. Mathematically, the terms describe the input and output values of a function, involving no causality beyond the association of each input value with a single output value. Mathematically, that's the end of the story; there's no deeper mathematical meaning in these terms.

However, the terms *independent* and *dependent* prompt students to think beyond the strict mathematical meaning and consider real-world causality. Can they change the

thing being measured by the independent variable, and in doing so cause a change in the thing being measured by the dependent variable? Such a physical (or geometric) dependence is more complex, more ambiguous, and more interesting than the precise and limited mathematical connection between independent and dependent variables.

Encourage students to argue about which variable should be independent and which dependent for particular combinations of values. Arguments will relate mainly to the non-mathematical meanings of the words. Point this out, and ask students to keep in mind also the mathematical meanings. They should end up with a sense of the distinction between the mathematical and non-mathematical meanings of these terms. In the mathematical realm, they should realize that the question of whether a particular pair of quantities can be considered independent and dependent variables is precisely the question of whether the second is a function of the first: Does the second measurement ever have more than one value corresponding to any particular value of the first?

The circle model on page 1 is a good place to start this discussion. Ask students whether the radius really depends on the diameter, or the diameter on the radius. The relation between the radius and the diameter is intentionally ambiguous; you can drag either and the other changes. You will likely have students willing to expound on either side of the argument, as well as some who will claim that both are correct. Also discuss the function relating the circumference to the radius, and challenge students to describe physical scenarios in which the circumference (or even the area) can be varied independently, with the radius depending upon it.

Another topic for the discussion is linear and nonlinear relationships. When do you expect the graph to be a line, and when do you expect it to be a curve? You might expect a line when plotting lengths against other lengths—such as a side length and the perimeter of a figure—and a curve when plotting a length against an area, but there are plenty of exceptions for students to encounter and explore.

CONSTRUCTION TIPS

Students will want to control the appearance and scale of their graphs. Show them how to do each of these tasks:

- Drag the origin using the **Arrow** tool.
- Hide the grid by choosing **Graph | Hide Grid.**
- Change the scale by dragging the tick mark numbers.
- Allow independent scaling of each axis by choosing **Graph | Grid Form | Rectangular.**

RADIUS AND DIAMETER

Q1 The diameter depends on the radius: Given a radius, there's only one possible value for the diameter.

Q2 It works both ways: Given a diameter, there's only one possible value for the radius. Both the relation and its inverse are functions, because if you know either of the two values, you also know the other.

Q3 Students can use either measurement in either role.

Q4 The graphs that students predict will vary. The crucial point is that they make a prediction ahead of time.

Q5 The graph shows direct variation with a slope of 2 or 0.5, depending on the choice of independent variable. The equations can be written *radius = diameter*/2 or *diameter* = 2 · *radius.*

Q6 Either measurement can be the independent variable. Both graphs are linear, but they have different slopes.

Q7 Both relations are functions, because there's only a single output value possible for any input value.

INVESTIGATE

Circle: A relation of the area and any linear measure results in half of a parabola, opening up or to the right depending on which is used for *x* and which for *y*. Because the measures are positive, these are all functions.

Triangle: (*AD, DE*) is direct variation (a linear function through the origin). The constant of proportionality (the slope) depends on the triangle's shape. For the original shape, the constant is approximately 0.92.

(*Distance A to DE, DE*) is also direct variation, with a constant that depends on the shape of the triangle.

Xquad: (*AD, DE*) is a function; the graph is in the shape of a V because it involves an absolute value. (*DE, AD*) is not a function: There are two possible values for *AD* that correspond to any given distance *DE*.

Rect 1: (*AD, AB*) is a function. *ABCD* is a constant-area rectangle, so the two sides are related by inverse variation, and the graph is in the shape of a hyperbola. Here's one form of the equation: $AB = (10/AD)$. This is one graph that's unchanged when students switch the dependent and independent variables.

(*AD, Perimeter ABCD*) is also a function, though (*Perimeter ABCD, AD*) is not. Here's the equation for the perimeter: $Perimeter\ ABCD = 2(10/AD) + 2AD$.

(*AD, Area ABCD*) is a constant function. $Area\ ABCD = 10$.

Rect 2: (*AD, AB*) is a linear function with a slope of -1. This is a constant-perimeter rectangle, so the sum of the lengths of the two sides stays constant: $AB = -AD + 10$.

(*AD, Perimeter ABCD*) is a constant function. The value of the constant depends on the length of a hidden segment in the construction.

(*AD, Area ABCD*) is a function, with a graph in the shape of a parabola opening down. (*Area ABCD, AD*) is not a function, because any value of the area (except when *ABCD* is a square) corresponds to two possible positions of *D*.

Polygons: For each polygon, the area is a function of side length, and the graph is in the shape of a half-parabola in Quadrant I. The graph is of form $Area = a\ AD^2$, where the value of *a* depends on the number of sides. For a triangle, $a = 0.5$; for a square, $a = 1$; for a pentagon,

$$a = \frac{5\varphi^2}{4\sqrt{\varphi^2 + 1}} = \frac{\sqrt{25 + 10\sqrt{5}}}{4} \approx 1.72$$

where φ is the golden ratio.

The perimeter of each polygon varies directly with the side length: $Perimeter = n \cdot AD$, where *n* is the number of sides.

Ladder: The relation (*AD, AB*) is a function. The graph is a quarter circle centered at the origin: $AB = \sqrt{10^2 - AD^2}$

Hanger: Because the four sides of the hanger are fixed in length, there are a number of constant functions here. The area is a function of *AD* and can be expressed as the square root of a quartic, related to Hero's formula.

Tube: The *field of view* varies directly with *distance*, and also varies directly with *tube diameter.* It varies inversely with *tube length.*

3

Systems

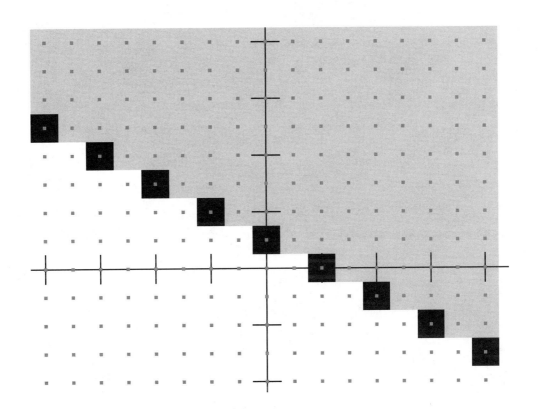

Solving Systems of Equations

The graphical solution to a system of two equations is the intersection of the graphs of the equations. In real life, you can often solve a system of equations to make a calculated choice between two options. In this activity you'll solve a system of equations to make some decisions for a truck rental company.

SKETCH

The end of the month is a popular time to move, so many people rent moving trucks. Each truck rental company has its own formula to determine the price of renting a truck. The rental rates depend on two factors: the *Drive-off Fee* and the *Price per Mile Driven*. Use the table below to compare the rates of two different companies.

Rate Comparison for 14-foot Moving Trucks

Company	Drive-off Fee	Price per Mile Driven
Cercano Rentals	$12.00	$2.75
We-Haul-It	$42.50	$0.40

With the rate information, the renter can make an informed decision. The price of renting a 14-foot truck will depend on the number of miles driven. Here is the price formula for Cercano Rentals:

$$f(x) = 12.00 + 2.75x$$

Q1 What is the price formula for We-Haul-It rentals?

Which company is cheaper depends on how many miles you drive. To compare the prices, you can graph each equation and find the intersection of the graphs.

The sketch is blank, but this document contains some tools that will be useful to you.

1. Open **Solving Systems.gsp** and choose **Graph | Define Coordinate System.** Hide the point at the origin and the point at $(1, 0)$. Use the **Arrow** tool to drag one of the numbers along the *x*-axis closer to the origin until you see "20, 40, 60, ..." for the scale on the *x*-axis. Then drag either axis to move the origin to the lower left-hand corner of the screen.

You can adjust the window, the scroll bars, and the scale of the *x*-axis to improve your view of the graph.

2. Choose **Graph | Plot New Function.** Type 12.00 + 2.75*x and click OK. The graph and its equation appear on the screen. With the graph and equation selected, choose **Display | Color** and make them blue.

3. Make sure no objects are selected. Choose **Graph | Plot New Function.** Type 42.50 + 0.40*x and click OK. The graph and its equation appear on the screen. Use **Display | Color** to make them dark green.

Solve this system of equations by finding the intersection of the graphs. First find an approximate solution.

4. Click the **Point** tool on one of the graphs. A point appears on the graph.

5. Use the **Arrow** tool to drag the point to the approximate intersection of the two lines. Choose **Measure | Coordinates.**

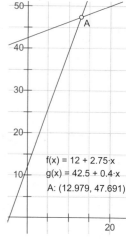

Q2 What are the values and units of the *x*- and *y*-coordinates that appear?

6. Now find a more exact location of the intersection. Drag the point away from the intersection to avoid future confusion.

7. Press and hold the **Custom** tools icon. From the menu that appears, choose the **Find Intersection** tool.

8. Click this tool on the two functions $f(x)$ and $g(x)$ in the sketch. (Be sure to click the functions themselves, not the function plots.) Then click a spot on the *x*-axis to the left of the intersection point, and finally a spot on the *x*-axis to the right of the intersection point.

The tool constructs an intersection of the two functions between the *x*-values of the two points and shows the coordinates of the intersection point.

Q3 What are the coordinates of the point of intersection? What are the units for each coordinate? What is the significance of this point?

Q4 Why do you think the point of intersection is often called the *break-even point*?

Q5 If the renter of the moving truck will drive 10 miles, from which company should he rent?

Q6 In Spanish, *cercano* means "nearby." Do you think that the name Cercano Rentals is appropriate for the way people use its trucks?

EXPLORE VARIATIONS

The owners of Cercano Rentals would like the company to be more competitive in moves of longer distances. The company investigates how it can change its rates to attract more long-distance movers. For marketing reasons, the company believes it is important to maintain its low *Drive-off Fee* of $12.00. Therefore, it decides to explore different scenarios by adjusting its *Price per Mile Driven*. Create a slider to make the *Price per Mile Driven* a variable rate.

9. Choose the **Basic Horizontal** tool and click in a blank space in the sketch to construct a slider.

To change the measurement's label, double-click it with the **Text** tool.

10. Change the label of the slider's measurement to *r*. Also change the label of the slider's active endpoint to match. The label *r* will represent the varying *Price per Mile Driven*. Drag the slider's endpoint back and forth to see how the measurement behaves.

11. Choose **Graph | Plot New Function.** Type 12.00 + r*x and then OK. (Click *r* in the sketch to insert it into the formula.). The graph and its equation appear on the screen. Choose **Display | Color** and make the line red.

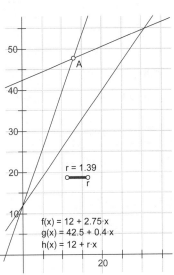

As you drag the point at the end of the slider to the left or right, you will notice that the slope of the red line changes and, as a result, so does the break-even point.

Use the **Find Intersection** tool to get accurate results to these questions.

Q7 Approximately what *Price per Mile Driven* should Cercano Rentals charge so that it is more competitive for moves of less than 40 miles? Less than 50 miles? Less than 60 miles? Less than 100 miles?

AN INTERNATIONAL QUANDARY

Add a page to the sketch by choosing **File | Document Options.**

Diego recently moved to California from Argentina and enjoys calling his family and friends back home to stay in touch. Currently, Diego pays $0.69 per minute for his calls to Argentina. The local phone company is encouraging Diego to buy an international calling plan. If he buys the plan, he will pay $3.99 per month plus an additional $0.15 per minute for his calls to Argentina.

Q8 On a new page, create and solve the system of equations to determine the minimum number of minutes of phone calls per month for which it would make sense for Diego to buy the international calling plan. Write down the equations you used and the break-even point.

EXPLORE MORE

Q9 If one or more of the equations in the system are non-linear, is there a solution to the system? Can there be more than one intersection point?

Q10 Can you write a system of equations that has two solutions? If yes, try it out.

Solving Systems of Equations

Objective: Students find the intersection of two graphs to solve a system of equations that reflects a real-world scenario. They then use a slider to see how the solution changes as the situation changes.

Student Audience: Algebra 1/Algebra 2

Prerequisites: Students should understand how to graph linear equations. It's best if they've already been introduced to systems of equations.

Sketchpad Level: Intermediate. Students plot functions on the graph and use two custom tools.

Activity Time: 30–40 minutes

Setting: Paired/Individual Activity (use **Solving Systems.gsp**) or Whole-Class Presentation (use **Solving Systems Present.gsp**)

SKETCH

Q1 The price formula for We-Haul-It is

$$g(x) = 42.50 + 0.40x$$

Q2 Answers will vary for the values, but should be close to (13, 47.7). The units are miles and dollars, so students might express the result as (13 miles, $47.70). Remind students that the y-value should be rounded to two decimal places because it is currency.

Q3 The numerical values to three decimal places are (12.979, 47.691). With the units, and rounding off the y-value to the nearest cent, the solution can be written (12.979 miles, $47.69). This point represents the distance at which the two rental companies have exactly the same cost.

Q4 It is called the *break-even point* because it is the place where the price to rent a truck is the same for both companies. The renter will neither lose nor save money by using the other company.

Q5 He should rent from Cercano Rentals because it will be cheaper.

Q6 Yes, Cercano Rentals is good for moves that are nearby, because it is cheaper for moves of less than about 13 miles.

Q7 The technique of using a slider to answer what-if questions is a powerful one. It's a good idea to talk about this technique during the wrap-up discussion.

To match We-Haul-It for a 40-mile trip, Cercano needs to set its price per mile at $1.16. Here are the rates Cercano must use to match We-Haul-It at other distances:

50 mi: $1.01   60 mi: $0.91   100 mi: $0.70

Note: Student answers may vary by a few cents because of limitations in manipulating the slider.

In case the intersection goes off-screen, remind students that they can rescale the axes by dragging the tick numbers.

Q8 This is the system of equations:
Current phone plan: $f(x) = 0.69x$
Calling card plan: $g(x) = 3.99 + 0.15x$
The point of intersection is (7.389 minutes, $5.09). If Diego talks for more than about seven minutes a month, the international calling plan will save him money.

Q9 Yes, there can still be a solution to a system of equations even if not all of the equations are linear.

Yes, there can be more than one intersection point.

Q10 Answers will vary. A simple example to show is a parabola with any linear equation that intersects it. For example, the system $f(x) = x^2 - 6x + 9$ and $g(x) = 0.5x + 1$ has two solutions. If you plot these two functions, you can use the **Find Intersection** tool to determine the actual intersections.

VARIATIONS

You may wish to give students different types of real-life word problems that involve finding an intersection. You could also ask them to make up their own break-even point problem.

WHOLE-CLASS PRESENTATION

You can present this activity to the entire class using **Solving Systems Present.gsp**.

Use the buttons and directions there to guide the presentation, and be sure to solicit student ideas and encourage discussion as you present.

Graphing Inequalities in Two Variables

Imagine yourself as a point in the coordinate plane, free to wander. In this activity you'll travel to different locations, learning how to keep inequalities satisfied.

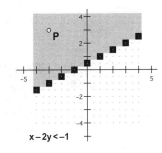

$x - 2y < -1$

KEEP ME SATISFIED

1. Open **Graphing Inequalities.gsp.** Move point P around the plane, and observe how the coordinates of the point vary.

Q1 Where can you move the point so that the calculation $2x + 3$ stays greater than 5? Describe these locations.

To edit the calculation, double-click it. Click the y measurement in the sketch to enter it into the calculation.

Q2 Edit the calculation to $2y + 3$. Now where can you move the point so that the calculation of $2y + 3$ stays greater than 5? Describe these locations.

Q3 Press *Show Tracer* and move point P around again, this time anywhere you like. What happens to the tracer when the calculation $2y + 3$ is greater than 5, when it's equal to 5, and when it's less than 5?

Although the inequality $2y + 3 > 5$ has only one variable, we're still interested in all the coordinate points, which have two variables.

A graph of an inequality in two variables is the collection of all the points in the plane that *satisfy* the inequality. In this context, satisfy means "make true." For example, any time point P is in a position where $2y + 3$ is greater than 5, the inequality $2y + 3 > 5$ is true.

Q4 What points satisfy the equation $2y + 3 = 5$? How might knowing this help you describe the points that satisfy the inequality $2y + 3 > 5$?

EXPLORE

To change the calculation, double-click it and then use the Calculator. Enter values of x or y by clicking the x or y measurements in the sketch.

2. Erase the traces, hide the tracer, and change the calculation to $3x - 2y$.

Q5 Given any position of point P, if you move it straight down (without moving it left or right), the value of $3x - 2y$ increases. Explain why. What happens to the calculation if you move P to the right without moving it up or down? Why?

3. Show the tracer and move point P around the plane.

Q6 Describe the locations of point P where $3x - 2y > 5$ is true. Use your answer to Q4 to explain why any point to the right of (or below) the line $3x - 2y = 5$ is a point that satisfies the inequality $3x - 2y > 5$.

4. On page 2, drag point P and notice how it moves between closely spaced points. Show the tracer and drag P again. The tracer changes color, showing for each location of point P whether $x + 2y > -1$ is true.

Q7 Use your answers to Q5 and Q6 to explain why one side of the line $x + 2y = -1$ contains *all* the points where $x + 2y > -1$.

Q8 Without using Sketchpad, use your answers to Q4 through Q6 to predict what points in the plane will satisfy the inequality $3x - y > 2$. Write out your thinking in a sentence, and draw a sketch of the graph on your paper. Then check your answer by editing the calculation and moving point P.

Q9 For each inequality below, sketch on your paper your prediction of the graph. Then use Sketchpad to check your result.

    a. $-x + 2y > -1$ b. $x + 2y > 1$

    c. $-x + 2y > 1$ d. $x - 2y > 1$

Q10 Use page 3 to find points that satisfy the inequality $y < x^2$. How can you determine, without Sketchpad, which points in the plane satisfy $y < x^2$?

Q11 Answer Q10 for the inequality $y > 3x - 5$.

EXPLORE MORE

To see the entire graph quickly, select both P and the tracer square, and choose **Construct | Locus.** Then use **Edit | Properties | Plot** to increase the number of samples so that the locus fills the entire region.

Q12 Use page 4 to find the points in the plane that satisfy the inequalities below. Draw a sketch of each solution. Think about how you could go about predicting the points that satisfy the inequality without Sketchpad available.

    a. $x^3 + 4 > 12$ b. $x^3 + 4 < 12$

    c. $x^3 + 4 < -12$ d. $|x| - |y| < 3$

    e. $|x| - |y| < -3$ f. $|x| - |y| > 3$

Q13 When you're done, compare the solutions to a, b, and c with each other, and the solutions to d, e, and f with each other. What conclusions, if any, can you draw about reversing the inequality sign?

Q14 Use page 5 to find the points that satisfy the inequalities below. How could you predict the points that satisfy the inequality without Sketchpad available?

    a. $(x - 2)(x + 1) < 0$ b. $(x - 2)(x + 1)(x - 4) < 0$

Q15 Use page 6 to find points that satisfy the inequality $3x - xy^2 + y^3 < 5$. (Make sure you drag point P to a variety of locations.) What points would satisfy the inequality $3x - xy^2 + y^3 > 5$? Why? What points would satisfy the inequality $-(3x - xy^2 + y^3) < -5$? Why? Check your answers with the sketch and explain the results that you see.

Graphing Inequalities in Two Variables

Objective: Students explore the graphs of inequalities by dragging a point on the Cartesian plane and observing where the inequality is true and where it is false. Students investigate the relationship between an inequality and the corresponding equation, and compare the graphs of different (but related) inequalities.

Student Audience: Algebra 1/Algebra 2

Prerequisites: None

Sketchpad Level: Easy

Activity Time: 40–50 minutes

Setting: Paired/Individual Activity (use **Graphing Inequalities.gsp**) or Whole-Class Presentation (use **Graphing Inequalities Present.gsp**)

Related Activities: Graphing Systems of Inequalities

KEEP ME SATISFIED

Q1 The calculation $2x + 3$ stays greater than 5 as long as P is to the right of a vertical line through 1 on the x-axis.

Q2 The calculation $2y + 3$ stays greater than 5 as long as P is above a horizontal line through 1 on the y-axis.

Q3 The tracer is cyan (light blue) when the calculation is greater than 5, black when it's equal to 5, and magenta (light purple) when it's less than 5.

Q4 The graph of $2y + 3 = 5$ is the horizontal line with a y-intercept of 1. The points satisfying $2y + 3 > 5$ are all points above that line.

EXPLORE

Q5 When you move P straight down, the y-value decreases. In the expression, y is multiplied by -2, so the value of the $-2y$ term increases when y decreases. Similarly, when you move P to the right, the x-value increases, so the value of the $3x$ term also increases.

Q6 The statement $3x - 2y > 5$ is true when P is to the right of or below the line that is the graph of $3x - 2y = 5$. Because the value of the expression is 5 when P is on the line, and the value increases when you move P to the right or down, every point on this side of the line must correspond to an expression value greater than 5.

Q7 In the expression $x + 2y$, the value of the expression increases when you move P to the right or up, because the first term (x) increases when you increase x by moving P right, and the second term ($2y$) increases when you increase y by moving P up. For this reason, the statement $x + 2y > -1$ is true when P is to the right of or above the line that is the graph of $x + 2y = -1$.

Q8 To satisfy the inequality $3x - y > 2$, the value of the expression must be greater than the value of the expression on the line $3x - y = 2$. The x term has a positive coefficient, so P must be to the right of the line; the y term has a negative coefficient, so P must be below the line. All points to the right of and below the line satisfy the inequality.

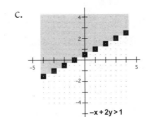

Q9   a.

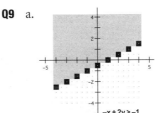

b.

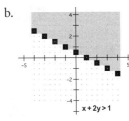

c.

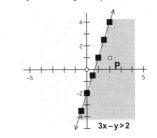

d.

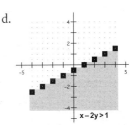

Q10 The points that satisfy the inequality $y < x^2$ are the points below the parabola $y = x^2$.

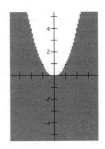

Q11 The inequality $y > 3x - 5$ is true when y is above the line because y has a positive coefficient and is on the greater-than side of the inequality. It's true when x is left of the line because x has a positive coefficient and is on the less-than side of the inequality.

EXPLORE MORE

Q12 a.

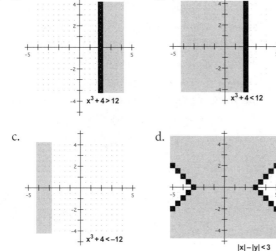

$x^3 + 4 > 12$

b.

$x^3 + 4 < 12$

c.

$x^3 + 4 < -12$

d.

$|x| - |y| < 3$

e.

$|x| - |y| < -3$

f.

$|x| - |y| > 3$

Q13 After students sketch the graphs, comparing graphs a and b and graphs d and f helps them to see that the direction of the inequality sign determines whether the cyan region or the magenta region represents the solution.

The curve(s) representing equality divide the plane into two regions. If you reverse the sign of the inequality, you change which region is the solution.

Q14 Each of these inequalities has a factored polynomial on the left side and zero on the right. To determine whether the inequality is satisfied for a given point, you only need to know the sign of the left side. Find that by determining the sign of each factor. The boundaries are vertical lines at the roots. Every time you cross one, one of the factors changes sign, and that changes the sign of the whole polynomial. Watch out for double roots; none are used in these examples.

Q15 This inequality has an interesting shape. Students may want to adjust the grid extent to view more of it.

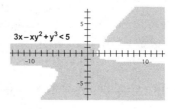

$3x - xy^2 + y^3 < 5$

The related inequality $3x - xy^2 + y^3 > 5$ has the inequality sign reversed, so the solution is the complementary region (the region not colored in the diagram). The inequality $-(3x - xy^2 + y^3) < -5$ has the same shape as $3x - xy^2 + y^3 > 5$, because the direction of the inequality is reversed, and the signs of both sides have also been reversed.

If you change the inequality to an equation and solve for x in terms of y, you can graph the result. (In the Function Calculator, set the Equation form to be $x = f(y)$.) You'll need to increase the number of samples in the function plot to see the complete graph.

WHOLE-CLASS PRESENTATION

Use the sketch **Graphing Inequalities Present.gsp** to present this activity to the class. The presentation follows the steps and questions of the student activity, but is streamlined to make it easy to present. Each page of the presentation sketch has directions and buttons to help you present the activity. Use pages 1 through 7 to present the main part of the activity. Pages 8 through 10 correspond to the Explore More questions.

Exploring Algebra 2 with The Geometer's Sketchpad
© 2007 Key Curriculum Press

Graphing Systems of Inequalities

When you solve a system of equations graphically, you usually get a single point as the solution. In this activity you'll solve a system of inequalities graphically, and see what the solution of such a system looks like.

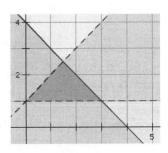

A SINGLE INEQUALITY

Begin by graphing the equation $y = \frac{3}{2}x - 2$ and the related inequality $y < \frac{3}{2}x - 2$.

1. In a new sketch, graph $y = \frac{3}{2}x - 2$ by choosing **Graph | Plot New Function** and entering the equation into the New Function dialog box.

Q1 What is the slope of the line and what are its x- and y-intercepts?

Q2 Based on the graph of the *equation*, what do you think the graph of the *inequality* will look like?

To delete the graph, select it and press the Delete or Backspace key on your keyboard.

2. Delete the function plot you just created, but leave the function itself; you will use it again in a moment.

To graph the inequality, you will use a custom tool from **Inequality Tools.gsp.**

3. Open **Inequality Tools.gsp.** Then switch back to your original sketch. (As long as **Inequality Tools.gsp** is open, you can use the tools it contains.)

Switch back to the **Arrow** tool after the graph appears.

4. Press and hold the **Custom** tools icon, and choose **Inequality Tools | y < f(x)** from the menu that appears. Click the function $f(x)$ in your sketch. A graph of the inequality appears.

You can also right-click (Windows) or control-click (Mac) on the strips to get to the Properties dialog box.

5. If the graph of the inequality appears as strips, you can change the Plot Properties of this object so that it appears as a solid area. Select the strips, choose **Edit | Properties,** and use the Plot panel of the Properties dialog box to increase the number of samples. (Approximately 400 should be enough.)

Q3 Is the boundary of the inequality graph solid or dashed? Explain why its appearance makes sense in terms of the inequality.

SYSTEMS OF INEQUALITIES

Next you'll graph a second inequality $(x \geq -y + 1)$ on the same coordinate system.

This function should be labeled g. If not, choose **Display | Label Function** to change its label to g.

6. Create another function by choosing **Graph | New Function.** This inequality expresses x in terms of y, so use the Calculator's Equation pop-up menu to choose **x = f(y).** Then enter the equation $g(x) = -y + 1$.

Q4 What do you think the graph of $x \geq -y + 1$ will look like? Write down your guess before you construct the graph.

Remember to switch back to the **Arrow** tool once the graph appears.

7. To create the graph, choose the custom tool **Inequality Tools | x >= f(y)**. Click the tool on the function $g(y)$ in your sketch.

Q5 Which shaded area contains points that satisfy both inequalities? Construct a point in that area, measure its coordinates, and use algebra to confirm that the point satisfies both inequalities.

In addition to the colors in the color menu, you can choose **Other...** to specify a different shade or tint.

Q6 For each system of inequalities below, add a new page to your document (using **File | Document Options**) and construct a graph of the system. If the graphs of two inequalities appear in the same color, change the color of one of the graphs so you can easily see the area of overlap.

a. $y \geq 2x - 1, y < -\frac{1}{2}x + 3$ b. $y < 2, x \geq -3$

c. $y \geq \frac{1}{2}x^2, x < y$ d. $y < \frac{1}{2}x^2, x \geq y - 1$

For problem h, use **Edit | Preferences | Units** and change the angle units to radians.

e. $x < y - 3, y \leq x - 1$ f. $y > \frac{x}{10} - x^2 + 3, y < 2$

g. $x \leq \sqrt{9 - y^2}, y > 1, x \geq -1$ h. $y < \sin x, y \geq 0.5$

EXPLORE MORE

Q7 Change these inequalities so that each variable appears on only one side of the inequality, making sure to keep the properties of inequality in mind. Then graph the system and describe the solution.

a. $2y - x < 2, |2x| \geq 2y - x$ b. $y^2 \leq x + 2y - 2, x - 3 < 0$

Some inequalities can be expressed with either x or y on the left side, so more than one answer may be possible for a particular graph.

Q8 Study each graph below and decide on a system of inequalities that will produce it. Check each answer on a new page of your document.

a. 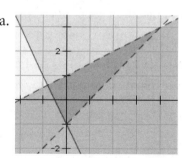   b.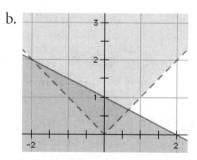

Q9 You probably expressed your answers to the preceding question in a form with y alone on the left side of each inequality. Where possible, rewrite your answers so that x is alone on the left side.

Q10 Use inequalities to graph an interesting shape in Sketchpad. Then challenge a friend to guess the inequalities you used.

© 2007 Key Curriculum Press

Graphing Systems of Inequalities

Objective: Students graph a single inequality and combine it with a second inequality to form the graph of a system of inequalities. They then graph several more such systems.

Student Audience: Algebra 2

Prerequisites: None

Sketchpad Level: Intermediate. This activity is easier if students have already used custom tools.

Activity Time: 35–45 minutes. To shorten the time, assign only selected problems in Q6 rather than all the problems.

Setting: Paired/Individual Activity (no sketch required, but uses tools from **Inequality Tools.gsp**) or Whole-Class Presentation (use **Systems of Inequalities Present.gsp**)

Related Activity: Graphing Inequalities in Two Variables

A SINGLE INEQUALITY

The graphing tools used in this activity require inequalities (like $y > x - 2$ or $x \le y^2 - 5$) that have been "solved" for either x or y. Most of the inequalities in this activity are already in this form. In Q7 students are asked to do the "solving" themselves. Use Q7 as a chance to review the properties of inequality needed to convert complicated expressions to the form required by the tools.

Q1 The graph crosses the y-axis at -2 and the x-axis at 1.33 and goes up and to the right with a slope of 1.5.

Q2 The inequality says that y must be less than the value of the function just graphed, so the graph of the inequality must be below the line. It extends infinitely down and to the right.

5. To avoid the need to increase the number of samples for every graph, hold down the Shift key and choose **Edit | Advanced Preferences**. On the Locus panel, change the number of point-locus samples from 300 to about 1000. (The exact value required depends on the size of the sketch window.)

Q3 The line itself is not included in the graph because y cannot be equal to the value of the function. For this reason it should appear as a dashed line.

SYSTEMS OF INEQUALITIES

Q4 Answers will vary; the important thing is that students make a conjecture before constructing the graph. (It's actually the region of the plane above and to the right of the line $x = -y + 1$.)

Q5 The graph is a portion of the plane bounded by the two lines, shown by the darkest area in the figure. This graph is infinite in extent.

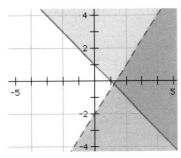

Q6 The area where the two shades overlap is the solution to the system. The graphs in problem e do not overlap, so it has no solution. In problem h, have students set the angle units to radians before graphing.

a.
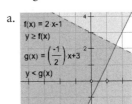

b.
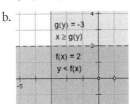

c.
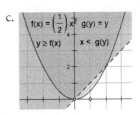

d.
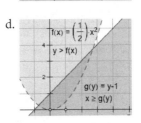

e.
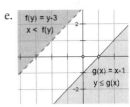

f.
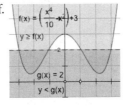

g.
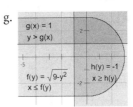

h.
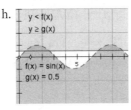

EXPLORE MORE

Q7 a. The first inequality can be expressed as $x > 2y - 2$ or as $y < 1/2x + 1$. The second inequality can be expressed as $y \geq x/2 + |x|$.

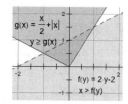

b. The first inequality can be expressed as $x \geq y^2 - 2y + 2$. The second inequality can be expressed as $x < 3$.

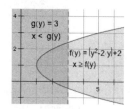

Q8 For some of the inequalities, it's possible to express the answers with either x or y on the left side. Here is one answer for each graph, with y on the left side of each inequality.

a. $y < 0.5x + 1$, $y > x - 1$, and $y \geq -2x - 1$

b. $y > |x|$ and $y \leq -0.5x + 1$

Q9 It's not possible to rewrite the inequality $y > |x|$ in a simple form with x on the left side. Here are the results for rewriting each of the other inequalities, with the original form on the left and the rewritten form on the right:

$$y < 0.5x + 1 \qquad x > 2y - 2$$
$$y > x - 1 \qquad x < y + 1$$
$$y \geq -2x - 1 \qquad x \geq -0.5y - 0.5$$
$$y \leq -0.5x + 1 \qquad x \leq -2y + 2$$

Q10 Answers will vary. Consider asking a few students to show and explain their answers to the class.

WHOLE-CLASS PRESENTATION

Use **Systems of Inequalities Present.gsp** to present this activity to the class. The presentation sketch first reviews graphing the single inequality $y \leq 3/2x - 2$.

1. On page 1, remind students that the first step in graphing an inequality is to graph the related equation. Use the button to show the function plot.

Q1 Ask students to predict where the graph of the inequality will be.

2. Hide the function plot and show the inequality.

Q2 Ask students why the boundary of the graph is a dashed line. Review the convention that a boundary that is part of the graph is shown as a solid line, and a boundary that is not part of the graph is shown dashed.

3. Go to page 2 to add the second inequality.

Q3 Ask students to predict where the graph will be. Have them explain how the graph of the inequality is related to the graph of $y = -x + 1$. Use Sketchpad's **Line** tool to draw an arbitrary line and have students tell you where to drag it to indicate the boundary.

Q4 Once students agree on where to place the boundary, ask them which side of the boundary the graph will be on. Use the tracer provided to check the result.

4. Press *Show y >= −x + 1* to show the result.

Q5 Ask students, "Why is this boundary solid?"

Q6 Ask students, "Where are the points that satisfy both inequalities? What is the solution region?"

5. Go to page 3, which contains a different system of inequalities.

Pages 3 through 8 contain various systems of inequalities for students to practice on. Use as many of these pages as are appropriate. Be sure to have students make a prediction before showing them the result on the computer.

Q7 Ask, "What will the first inequality look like?"

6. Use the buttons to show first the related function and then the inequality.

7. Similarly, ask about and show the second function and inequality, and then discuss the solution.

Pages 9 and 10 contain graphs of systems of inequalities. (These are the graphs from Q8 in the activity.) Use either page as a challenge to students to come up with a system of inequalities that matches the graph. Use the custom tools in **Inequality Tools.gsp** to verify student answers.

Linear Programming: Swans and Giraffes

In business, it's often difficult to figure out how to maximize profit because there are so many factors to consider: labor costs, time constraints, production capacity, etc. *Linear programming* is a branch of algebra that helps deal with complicated situations such as this.

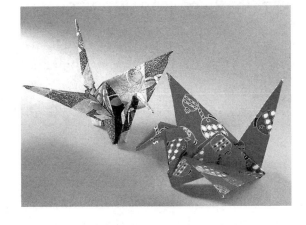

Here's a typical linear programming problem:

Origami is the ancient Japanese art of paper folding.

Rei volunteers to bring origami swans and giraffes to sell at a charity crafts fair. It takes her three minutes to make a swan and six minutes to make a giraffe. She plans to sell the swans for $4 each and the giraffes for $6 each. If she has only 16 pieces of origami paper and can't spend more than one hour folding, how many of each animal should Rei make to maximize the charity's profit?

ASSIGN VARIABLES AND WRITE THE EQUATION

There's a lot of information to deal with here. The actual question is "How many of each type of animal should Rei make?" so start by assigning variables:

Let s = the number of swans Rei makes.

Let g = the number of giraffes Rei makes.

In other words, if Rei makes s swans and g giraffes, how much money will that bring in?

Q1 Write an equation for profit in terms of s and g.

1. Open **Linear Programming.gsp.** You'll see the profit equation calculated using the coordinates of point R. Each location of point R is a potential solution to the problem. For example, (6, 8) represents a solution of six swans and eight giraffes. (Does the profit equation agree with the one you came up with?)

Q2 Drag point R around and observe how the coordinate and profit measurements change. Describe how the profit relates to R's location in the plane. Where is profit greatest? Where is it smallest?

2. Choose **Graph | Snap Points** to turn on point snapping. Drag R again.

These limitations are *implicit constraints:* constraints implied in the problem but not explicitly mentioned.

Q3 What is the effect of point snapping? Why does it make sense to use it for this problem? In addition to point snapping, what other constraints (limitations) should there be when you drag R? Why do your proposed constraints make sense in terms of the original problem?

SET UP THE CONSTRAINT INEQUALITIES

The more origami figures Rei makes, the greater the profit will be. But this ignores the *constraints* (limitations) listed in the problem. The first constraint is that she has only 16 pieces of origami paper. We'll call this the *paper constraint* and express it mathematically with the *constraint inequality* $s + g \leq 16$.

3. Go to page 2. You'll see a new line and a new calculation. The line $s + g = 16$ defines the boundary of the paper constraint, and divides the first quadrant into two regions. The calculation, $s_R + g_R$ shows how many pieces of paper are used for the current point R.

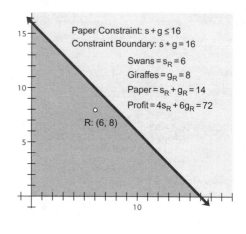

Q4 Where in the sketch can you drag R and have the paper constraint satisfied? Within that region, where is the profit greatest?

In the next two steps you'll develop a similar inequality for the *time constraint*.

Q5 The problem states: "It takes her three minutes to make a swan and six minutes to make a giraffe." Use this information to write two expressions: one for how long it takes Rei to make s swans, and the other for how long it takes her to make g giraffes.

Q6 Use your expressions from Q5 to write a constraint inequality for the time constraint—that Rei can't spend more than one hour folding.

4. Go to page 3. This page includes a blue line for the limit of the time constraint. This line and the red line divide the first quadrant into four regions, each with a different color.

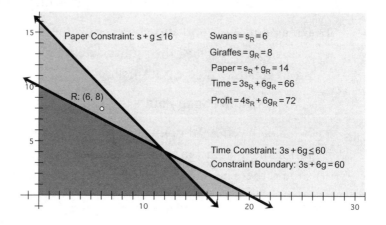

Q7 For each of the four regions, list which of the two major constraints (paper and time) are met and which aren't. In which region(s) can you drag *R* and have both major constraints met?

FIND THE MAXIMUM

The region you found in Q7 in which both constraints are satisfied is called the *feasible region.* Your goal now is to find the one point within this region that maximizes profit. You can drag point *R* around and try to find where the highest profit occurs, but there's an easier way. Linear programmers can prove that *the maximum value always occurs at a corner point of the feasible region.* But which corner is it?

5. To investigate, first attach point *R* to the feasible region. To do so, select both *R* and the feasible region, and choose **Edit | Merge Point To Quadrilateral.**

Q8 Drag point *R* around the perimeter of the feasible region, noting the profit value at the corner points. How many swans and giraffes should Rei make in order to maximize the profit? What is the maximum profit?

EXPLORE MORE

Q9 Suppose that Rei actually has 18 pieces of paper. Change page 3 of the sketch to investigate this new problem. (To do so, press the *Show Parameters* button, then double-click *Sheets of Paper* and change its value.) How does this change the feasible region? Can Rei increase profits with this new feasible region?

Q10 Continue experimenting with changing the paper constraint. How much does the maximum profit increase with each extra sheet?

Q11 Rei was limited by only two major constraints. What if a third were added? For instance, what if Rei didn't want to make more than eight swans? Turn this new constraint into a constraint inequality and describe the new feasible region as well as the maximum profit.

Q12 Go to page 4 of **Linear Programming.gsp.** You'll see a new line labeled *Profit.* Alternate between dragging point *R* along this new line and dragging the line itself. (Drag the line by dragging its *y*-intercept.) Explain what the new line represents. Then explain how it demonstrates why profit will always be maximized at a corner point of the feasible region.

Linear Programming: Swans and Giraffes

Objective: Students explore a linear programming problem. They turn the problem statements into constraint equations, explore the feasible region, and determine how to maximize a desired quantity.

Student Audience: Algebra 2

Prerequisites: Students should be fluent with linear equations, simultaneous equations, and word problems.

Sketchpad Level: Easy

Activity Time: 40–50 minutes

Setting: Paired/Individual Activity (use **Linear Programming.gsp**) or Whole-Class Presentation (use **Linear Programming Present.gsp**)

Related Activity: Graphing Systems of Inequalities

ASSIGN VARIABLES AND WRITE THE EQUATION

Q1 Profit $= 4s + 6g$

Q2 The closer point R is to the origin, the less the profit. The farther up and to the right, the greater the profit. (Some students may notice already that the relationship is a bit more subtle than this, because the profit at $(0, 16)$ is more than the profit at $(16, 0)$. This observation becomes important when answering Q4.)

Q3 It makes sense to have point R snap to locations with integer coordinates because Rei can sell only whole numbers of swans and giraffes. It doesn't make sense to consider her making 3.71 swans or 4.2 giraffes.

Besides limiting both s and g to integers, there are two more implicit restrictions: $s \geq 0$ and $g \geq 0$: Rei cannot make negative numbers of either animal. These restrictions limit R to the first quadrant.

SET UP THE CONSTRAINT INEQUALITIES

Q4 The paper constraint is satisfied only within (and on the perimeter of) the triangle bounded by the two axes and the purple line. Within this region, the profit is greatest at the upper-left corner, or $(0, 16)$. This makes sense since, without the time constraint, the greatest profit would be made by using all 16 pieces of paper to make giraffes.

Q5 $3s =$ how long it takes Rei to make s swans
$6g =$ how long it takes Rei to make g giraffes

Q6 $3s + 6g \leq 60$

Q7 In the yellow region, neither constraint is met. In the light green region, only the time constraint is met. In the red region, only the paper constraint is met. In the dark blue region, both are met.

FIND THE MAXIMUM

Q8 Rei should make 12 giraffes and 4 swans for a profit of $72.

EXPLORE MORE

Q9 This increases the size of the feasible region a bit, moving the upper-right vertex a little down and to the right. Rei would be able to make a greater profit with this change. In fact, increasing the number of pieces of paper Rei has would increase her profit up to the point at which she has 20 pieces. Having more than 20 pieces of paper wouldn't make any difference (because of the time constraint).

Q10 Each additional piece of paper increases her profit by $2, to a maximum of $80 profit with 20 pieces of paper.

Q11 The line $g \leq 8$ corresponds to the constraint "Rei didn't want to make more than eight swans." This is a region to the left of a vertical line that cuts the original feasible region into a smaller pentagonal region. The maximum profit for this situation is $68, occurring again at the upper-right vertex of the feasible region: eight swans and six giraffes.

Q12 The profit line represents all combinations of giraffes and swans that give the same profit. To see this, drag point R along the line: The profit doesn't change. With the profit line far up and to the right, you get a large profit, but R doesn't touch the feasible region. With the profit line cutting through the feasible region, R can take on feasible values, but the profit isn't at its greatest. Only when the profit line just touches a corner of the feasible region is the profit maximized.

4

Quadratic Functions

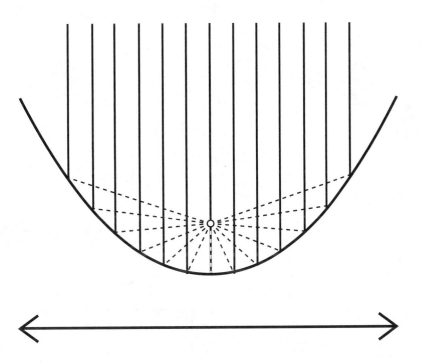

Parabolas in Vertex Form

Things with *bilateral symmetry*, such as the human body, have parts on the sides that come in pairs (such as ears and feet) and parts down the middle that there's just one of (such as the nose and belly button). Parabolas are the same way. Points on one side have corresponding points on the other. But one point is unique: the vertex. It's right in the middle, and, like your nose, there's just one of it. Not surprisingly, there's a common equation form for parabolas that relates to this unique point.

SKETCH AND INVESTIGATE

1. Open **Vertex Form.gsp.** You'll see an equation in the form $y = a(x - h)^2 + k$, with *a*, *h*, and *k* filled in, and sliders for *a*, *h*, and *k*. Adjust the sliders (by dragging the points at their tips) and watch the equation change accordingly. There's no graph yet; you'll create that in the next step.

To enter *a, h,* and *k*, click their measurements in the sketch. To enter *x*, click on the *x* in the Calculator keypad.

2. Choose **Graph | Plot New Function.** The New Function dialog box appears. If necessary, move it so that you can see the measurements of *a*, *h*, and *k*. Enter $a*(x-h)\wedge 2 + k$ and click OK. Sketchpad plots the function for the current values of *a*, *h*, and *k*.

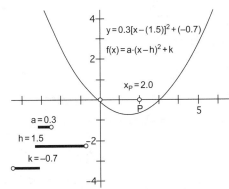

Notice point *P* on the *x*-axis. The measurement x_p is the *x*-coordinate of this point. You'll now plot on the parabola a point that has the same *x*-coordinate.

Choose **Measure | Calculate.** Click on the function equation from step 2. Then click on x_p to enter it, and click OK.

3. Calculate $f(x_p)$, the value of the function *f* evaluated at x_p. You'll see an equation for $f(x_p)$, the value of the function *f* evaluated at x_p.

4. Select, in order, x_p and $f(x_p)$. Choose **Graph | Plot As (x, y).** This command plots a point on the parabola. Label the new point *Q*. Drag *P* if you can't see *Q*.

Q1 Using paper and pencil or Sketchpad's Calculator, show that the coordinates of point *Q* satisfy the parabola's equation. If the numbers are a little off, explain why this might be.

EXPLORE FAMILIES OF PARABOLAS

By dragging point *P*, you're exploring how the variables *x* and *y* vary along one particular parabola with particular values for *a*, *h*, and *k*. For the rest of this activity, you'll change the values of *a*, *h*, and *k*, which will change the parabola itself, allowing you to explore whole families of parabolas.

Q2 Adjust slider *a* and observe the effect on the parabola.

Parabolas in Vertex Form

continued

Summarize the role of a in the equation $y = a(x - h)^2 + k$. Be sure to discuss its sign, its magnitude, and anything else that seems important.

Q3 Changing a appears to change all the points on the parabola but one, the vertex. Change the values of h and k; then adjust a again, focusing on where the vertex appears to be. How does the location of the vertex relate to the parameters h and k?

Q4 Adjust the sliders for h and k. Describe how changing h transforms the parabola. How does that compare to the way that changing k transforms it?

To use **Graph | Plot As (x, y),** first select exactly two measurements to be used as the values for x and y.

5. Plot the vertex of your parabola using the **Graph | Plot As (x, y)** command.

Q5 Write the equation in vertex form $y = a(x - h)^2 + k$ for each parabola described below. As a check, adjust the sliders so that the parabola is drawn on the screen.

Note: In this activity, the precision of measurements has been set to one decimal place. It's important to be aware of this and to check your answers by hand, in addition to adjusting the sliders in the sketch.

   a. vertex at $(1, -1)$; y-intercept at $(0, 4)$

   b. vertex at $(-4, -3)$; contains the point $(-2, -1)$

   c. vertex at $(5, 2)$; contains the point $(1, -6)$

   d. same vertex as the parabola $-3(x - 2)^2 - 2$; contains the point $(0, 6)$

   e. congruent to the parabola $4(x + 3)^2 - 1$; vertex at $(-1, 3)$

Q6 The axis of symmetry is the line over which a parabola can be flipped and still look the same. What is the equation of the axis of symmetry for the parabola $y = 2(x - 3)^2 + 1$? What is the general equation for the line of symmetry corresponding to the parabola $y = a(x - h)^2 + k$?

Q7 Just as your right ear has a corresponding ear across your body's axis of symmetry, all points on a parabola (except the vertex) have corresponding points across its axis of symmetry.

The point $(5, 9)$ is on the parabola $y = 2(x - 3)^2 + 1$. What is the corresponding point across the axis of symmetry?

EXPLORE MORE

Q8 Move point Q (s, t) to the right half of the parabola $y = a(x - h)^2 + k$. What is the corresponding point Q' across the axis of symmetry, in terms of s and t? If (s, t) were on the left half of the parabola, what would the answer be?

Q9 Use **Construct | Perpendicular Line** to construct the axis of symmetry of your parabola. Then use **Transform | Reflect** to reflect point Q across the axis of symmetry. Measure the coordinates of the new point, Q'. Are they what you expected?

Parabolas in Vertex Form

Objective: Students vary parameters on a quadratic function in vertex form and create a graph satisfying specific conditions.

Student Audience: Algebra 1/Algebra 2

Prerequisites: Students need to understand the basic idea of a function and the role that the variables x and y play in the equation and graph of a function. Solving simple linear equations for one unknown after substituting given values for other unknowns is also part of this activity.

Sketchpad Level: Easy

Activity Time: 40–50 minutes

Setting: Paired/Individual Activity (use **Vertex Form.gsp**) or Whole-Class Presentation (use **Vertex Form Present.gsp**)

Related Activities: The activity Exploring Parabolas in Vertex Form covers much of the same material, but in a more open-ended way.

You may wish to draw attention to the similarities between this equation form for the parabola and the point-slope form of a line (as well as the formula used in the Absolute Value Functions activity, if students have done that). A good topic of discussion is whether there's a notion of slope with parabolas. If so, how is it different from the slope of a line? If not, what does a control?

SKETCH AND INVESTIGATE

Q1 There are many possible answers.

This is a good calculator activity for substituting an x-value into the right side of a function to determine a y-value. Any small discrepancy is due to Sketchpad's rounding (depending on the Precision settings in Preferences).

EXPLORE FAMILIES OF PARABOLAS

Q2 If a is positive, the parabola opens upward; if a is negative, the parabola opens downward. The larger the absolute value of a, the narrower the parabola. The closer a is to zero, the wider the parabola.

Q3 The coordinates of the vertex are (h, k).

Q4 The parabola moves right and left as h changes (right as h gets bigger, left as it gets smaller). The parabola moves up and down as k changes (up as k gets bigger, down as it gets smaller).

Q5 a. $y = 5(x - 1)^2 - 1$

  b. $y = 0.5(x + 4)^2 - 3$

  c. $y = -0.5(x - 5)^2 + 2$

  d. $y = 2(x - 2)^2 - 2$

  e. $y = 4(x + 1)^2 + 3$

It's very important that students find the equations of these parabolas using paper-and-pencil calculations and use Sketchpad to check their answers. Slider accuracy may account for small differences.

Q6 $x = 3; x = h$

Q7 $(1, 9)$

EXPLORE MORE

Q8 Since $x = h$ is the axis of symmetry, $Q'(2h - s, t)$ is the reflected image of $Q(s, t)$. This relationship will hold no matter which side of the parabola point Q is on.

Q9 The coordinates of points Q and Q' should be consistent with the relationship described above.

Exploring Algebra 2 with The Geometer's Sketchpad
© 2007 Key Curriculum Press

4: Quadratic Functions | 75

Parabolas in Vertex Form

Before beginning this presentation, it would be a good idea to review the fact that graphs of quadratic functions are parabolas. Also, you can expand some vertex form quadratic functions to show that they are in fact quadratic.

1. Open **Vertex Form Present.gsp.** Drag the sliders one at a time so that students can see how they connect with the function at the top of the screen.

2. Press the *Show Graph* button.

Q1 Ask the class what will happen when you drag the slider for parameter *a*. After giving them sufficient time for discussion, drag the slider. Be sure to show positive and negative values, and give some attention to the special case of $a = 0$.

Q2 Move on to the sliders for parameters *h* and *k*. Again, ask students to make predictions before you actually change any values.

Q3 What are the coordinates of the vertex of this parabola? (The coordinates are (h, k).)

3. Select in order parameters *h* and *k*. Choose **Graph | Plot As (x, y).** Drag the sliders one at a time to show that the new plotted point stays at the vertex.

Q4 Ask students how we can change this parabola so that the vertex is at $(0, 4)$ and the graph contains the point $(1, -1)$. The most obvious way is to alternate dragging each of the three sliders until the desired result appears. Use whatever method the class suggests.

Q5 Hide the graph. Ask students how to put the vertex at $(-4, -3)$ and make the graph go through point $(-2, -1)$. Knowing the vertex should tell them that $h = -4$ and $k = -3$. Show them how to substitute these values into the equation along with $x = -2$ and $y = -1$. Then solve for a $(a = 0.5)$. Press *Show Graph* to test your solution. Try a few other examples.

Q6 All parabolas have an axis of symmetry. Where is the axis of this parabola? What is the equation of the axis? (The axis is a vertical line through the vertex, and its equation is $x = h$.)

4. Select the vertex and the *x*-axis. Choose **Construct | Perpendicular Line.**

5. Construct a point on the parabola.

Q7 Ask, "If we reflect this point across the axis of symmetry, where will it fall?" (The reflected image should also be on the parabola.)

6. Double-click the axis of symmetry. Select the point on the parabola. Choose **Transform | Reflect.** Drag both the point and the sliders to show that this works in all cases.

Exploring Algebra 2 with The Geometer's Sketchpad
© 2007 Key Curriculum Press

Exploring Parabolas in Vertex Form

This activity is an open-ended exploration of quadratic functions expressed in vertex form. Play with the various pages in **Explore Vertex Form.gsp** and let your curiosity be your guide. The questions below are meant to point you toward interesting areas of inquiry, but you or your teacher may decide that other questions are more interesting or relevant. Happy exploring!

EXPLORE

Open **Explore Vertex Form.gsp.** Use the link buttons to navigate from page to page. The various pages are described below.

$y = x^2$ (**basic parabola**)	The vanilla parabola, the base model: no frills, no coefficients, nothing added. Understand this parabola and you're on your way to understanding them all.
$y = ax^2$ (**slider**) $y = ax^2$ (**direct**)	What happens when you multiply x^2 by a constant? Find out on these two pages. On the first, the parameter a is controlled by a slider. On the second, it's controlled directly by a point on the parabola. The underlying math is the same for both.
$y = ax^2 + k$ (**slider**) $y = ax^2 + k$ (**direct**)	What happens when you add a constant k to ax^2? Again, a and k are controlled with sliders on one page and directly with points on the parabola on the other, but the math is the same.
$y = a(x - h)^2 + k$ (**slider**) $y = a(x - h)^2 + k$ (**direct**)	What happens when you now subtract a constant h from the x in $y = ax^2 + k$? Find out here.

QUESTIONS TO PONDER, DISCUSS, OR WRITE ABOUT

Q1 Name all of the points on $y = x^2$ that are visible in the current window and that have integers for both x- and y-coordinates. Without scrolling, name four other points on $y = x^2$ with integer coordinates.

Q2 What can you say about the equation of a parabola if its vertex is at the origin?

Q3 What kind of symmetry do all parabolas in the family $y = ax^2$ exhibit? Why do they have this symmetry?

Q4 If the point (c, d) is on the parabola $y = x^2$, name one other point that must be on the parabola.

Q5 How does the sign of a affect the parabolas on these pages?

Q6 What do the various graphs look like if $a = 0$?

Q7 How is the coefficient a in these equations similar to the coefficient m in $y = mx$, $y = mx + b$, and $y = m(x - h) + k$? How is it different?

Q8 The third page shows $y = ax^2$ and lets you control a with a drag control located one unit to the right of the axis of symmetry (the y-axis). Why does it make sense for the drag control to be there?

Q9 The axis of symmetry is the line a parabola can be flipped over and still look the same. The axis of symmetry for $y = x^2$ is $x = 0$ (the y-axis). What is the equation of the axis of symmetry for parabolas in the family $y = ax^2$? $y = ax^2 + k$? $y = a(x - h)^2 + k$?

Q10 Transformations used in geometry include translations, rotations, dilations, reflections, stretchs, shrinks, and shears. Look up these terms, then describe which apply to the transformations to parabolas you observe when dragging the sliders for a, h, and k.

EXPLORE MORE

Use commands from the Graph and/or Construct menus to construct the axis of symmetry of the parabola on one of the last two pages. Then use commands from the Transform menu to reflect point P across this axis.

Exploring Parabolas in Vertex Form

Objective: Students use a prepared sketch to explore the connections between parameters of quadratic functions and their graphs.

Student Audience: Algebra 1/Algebra 2/Geometry

Prerequisites: None in particular. Some familiarity with parabolas and with graphical transformations would help.

Sketchpad Level: Easy. Students manipulate sliders in a pre-made sketch.

Activity Time: 20–40 minutes. This depends on how many questions are explored, whether answers are written or discussed, how detailed the answers are, and so on.

Setting: Paired/Individual Activity (use **Explore Vertex Form.gsp**) or Whole-Class Presentation (use **Explore Vertex Form Present.gsp**)

Related Activities: The activity Parabolas in Vertex Form is concerned with the same subject matter as this activity, but approaches it in a much different way. That activity uses a more traditional, step-by-step approach, and it focuses exclusively on the vertex form. This activity uses a more open-ended approach and builds toward the vertex form from much more basic parabola forms. Which approach is more appropriate depends on the particular students and the particular class involved.

QUESTIONS TO PONDER, DISCUSS, OR WRITE ABOUT

Q1 The integer points visible in the sketch window are $(0, 0)$, $(1, 1)$, $(-1, 1)$, $(2, 4)$, $(-2, 4)$, $(3, 9)$, and $(-3, 9)$. Four other integer points not visible but on the parabola are $(4, 16)$, $(-4, 16)$, $(5, 25)$, and $(-5, 25)$. (Many other answers are possible for the second part of this question.)

Q2 It must be of the form $y = ax^2$. In vertex form, a parabola with vertex $(0, 0)$ is written as $y = a(x - 0)^2 + 0$, which simplifies to $y = ax^2$.

Q3 They are all symmetrical across the y-axis. This is because the square of a number is the same as the square of its opposite. For example, if $y = 2x^2$, both $(3, 18)$ and $(-3, 18)$ are on the graph because $3^2 = (-3)^2$. These two points are reflections of each other across the y-axis. Functions that have this property, $f(-x) = f(x)$, are called *even* functions.

Q4 Point $(-c, d)$ is also on the parabola.

Q5 If a is positive, the parabola opens upward. If a is negative, it opens downward. As the magnitude of a gets bigger, the parabola takes on narrower form.

Q6 They become horizontal lines.

Q7 Coefficient m controls the slopes of lines and a controls something for parabolas that's very much like slope: how narrow (steep) or wide (flat) a parabola is. The sign of m determines whether the graph rises or falls as you move to the right. The sign of a determines whether the graph rises or falls as you move horizontally away from the vertex.

Q8 It makes sense because the y-coordinate at $x = 1$ is a. To see this, just substitute 1 for x in $y = ax^2$. You get $y = a$.

Q9 The equation of the axis of symmetry for both $y = ax^2$ and $y = ax^2 + k$ is $x = 0$. The equation of the axis of symmetry for $y = a(x - h)^2 + k$ is $x = h$.

Q10 Dragging h and k results in horizontal and vertical translations, respectively. Dragging a stretches the graph vertically.

EXPLORE MORE

One way to do this is to construct a vertical line through the vertex. The vertex is already plotted on the direct-control page. To plot the vertex on the slider-control page, select h and k in that order and choose **Graph | Plot As (x, y)**. Now select the vertex and the x-axis and choose **Construct | Perpendicular Line.**

Parabolas in Factored Form

If you use a parabola to model a thrown ball, you might want to know exactly where it hits the ground. Or if you model a profit function with a parabola, you might be interested in the *break-even point*—the cutoff between profitability and loss. In both cases you're interested in the *roots* of the equation—where the function's value is zero. In this activity you'll explore a quadratic form that's based on the roots.

SKETCH AND INVESTIGATE

1. Open **Factored Form.gsp.** This sketch contains three sliders (a, r_1, and r_2). Adjust each slider by dragging the point at its tip. Observe how the values change as you drag.

$a = -0.5$

$r_1 = 2.0$

$r_2 = -1.7$

First you'll create a function, evaluate it for an input variable, and plot the resulting point.

2. Use the slider values to define the function $f(x) = a(x - r_1)(x - r_2)$. To do so, choose **Graph | New Function.** The New Function dialog box appears. If necessary, move it so that you can see the measurements of a, r_1, and r_2.

To enter a, r_1, and r_2, click their measurements in the sketch. To enter x, click the x key in the dialog box.

3. Enter $a*(x-r_1)*(x-r_2)$ and click OK. Sketchpad creates the function.

4. Measure the x-coordinate of point P by choosing **Measure | Abscissa.**

5. Calculate $f(x_p)$, the value of function f evaluated at x_p. To do so, choose **Measure | Calculate.** Click on the function object $f(x)$ and then on measurement x_p. Finally click OK. The value of the function appears.

6. To plot the point, select x_p and $f(x_p)$ and choose **Graph | Plot As (x, y).**

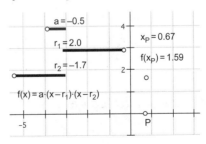

To turn tracing on or off, select the plotted point and choose **Display | Trace Plotted Point.**

Q1 Drag P back and forth along the x-axis to change the input variable for the function. How does the plotted point behave? Turn on tracing for the plotted point to better observe its behavior. What shape does this function trace out?

Traces are temporary in Sketchpad. Instead of using traces, you'll now make the graph permanent so you can compare the graphs of different functions.

7. Turn off tracing for the plotted point. Then choose **Display | Erase Traces** and drag *P* to make sure the traces no longer appear.

8. Plot a permanent graph of the function by selecting the function and choosing **Graph | Plot Function.** Drag *P* back and forth to make sure that the graph really corresponds to the path of the plotted point.

EXPLORING FAMILIES OF PARABOLAS

By dragging point *P*, you explored how *x* and *y* vary for *one particular function* with specific values of *a*, r_1, and r_2. Now you'll change the values of *a*, r_1, and r_2, which *changes the function itself,* allowing you to explore whole families of parabolas.

Q2 Adjust slider *a* and describe its effect on the parabola. Discuss the effect of *a*'s sign (whether it's positive or negative), its magnitude (how big or small it is), and anything else that seems important.

Q3 Dragging *a* appears to change all the points on the parabola but two: the *x*-intercepts of the parabola (the roots). Adjust all three sliders and observe the effect that each has on the *x*-intercepts. How are the locations of the *x*-intercepts related to the values of the sliders?

Q4 Adjust slider r_1. What happens to the parabola as r_1 changes? What happens as r_2 changes?

Q5 Adjust the sliders so that $r_1 = r_2$. Describe the resulting parabola.

Figure out the equations using pencil and paper only. Once you think you have the equation, you can use the sketch to check your result.

Q6 For each description below, write an equation for a parabola in factored form $f(x) = a(x - r_1)(x - r_2)$. Check your answers by adjusting the sliders.

a. *x*-intercepts at $(-4, 0)$ and $(6, 0)$; vertex at $(1, -1)$

b. *x*-intercepts at $(-5, 0)$ and $(1, 0)$; contains the point $(3, 32)$

c. *x*-intercepts at $(0, 0)$ and $(-3, 0)$; contains the point $(2, 3)$

d. same *x*-intercepts as $y = 2(x - 3)(x + 1)$; contains the point $(0, -3)$

e. same shape as $y = 2(x - 3)(x + 1)$; *x*-intercepts at $(-4, 0)$ and $(1, 0)$

Q7 You throw a baseball and it flies in a parabolic path across a field. If the ball reaches its apex (highest point) 60 feet away from you, and the apex is 40 feet above the ground, how far away from you will the ball land? What is the equation of the ball's flight in factored form? (Assume that the ball starts at the point $(0, 0)$.)

EXPLORE MORE

Q8 When you have a parabola in the form $f(x) = a(x - h)^2 + k$, it's easy to find its vertex, but harder to find its roots. The opposite is true with the form $f(x) = a(x - r_1)(x - r_2)$. Explain what you do know about the vertex of parabolas in this form. Can you write an expression for the x-coordinate of the vertex in terms of r_1 and r_2? The y-coordinate?

9. Use the expressions you just wrote for the coordinates of the vertex to plot the vertex. If you do this properly, the plotted point will remain at the vertex regardless of how you drag the sliders. You can use Sketchpad's Calculator to calculate the x- and y-coordinates of the vertex using the expressions you found in the previous question. You can then select the two calculations and choose **Graph | Plot As (x, y).**

Q9 If you successfully plotted the vertex, try this: Turn on tracing for the vertex by choosing **Display | Trace Plotted Point.** Then adjust slider r_1. What shape does the vertex trace? Can you write an equation for this curve in terms of a and r_2?

Parabolas in Factored Form

Objective: Students plot the graph of a quadratic function in factored form, investigate the relationship between this form and its graph, and use their observations to create functions from various descriptions of their graphs.

Student Audience: Algebra 1/Algebra 2

Prerequisites: Students should have some experience with graphing quadratic equations. It's not necessary for them to be familiar already with the factored form.

Sketchpad Level: Intermediate. Students create and evaluate a function, plot and trace a point, and plot a function.

Activity Time: 40–50 minutes. The activity is shorter if you use the alternate sketch **Factored Form 2.gsp** and start with Exploring Families of Parabolas on the second page of the activity.

Setting: Paired/Individual Activity (use **Factored Form.gsp**) or Whole-Class Presentation (use **Factored Form Present.gsp**)

Related Activities: Parabolas in Standard Form, Parabolas in Vertex Form, Changing Quadratic Function Forms

It's important that students understand that a parabola remains a parabola regardless of the form its equation is written in. Any parabola's equation can be written in many different forms. The forms we study are the ones that are especially simple or that convey special information.

The factored form of a parabola is different from the standard form and the vertex form, because parabolas that don't cross or touch the x-axis cannot be written in factored form. When you summarize this activity, you might discuss the existence of roots and the possibility of categorizing parabolas based on whether they have 0, 1, or 2 distinct real roots. If students have studied the discriminant, this would be a good connection to make.

SKETCH AND INVESTIGATE

Q1 The plotted point follows a curved, parabolic path. The exact shape (narrow or wide, opening up or down) depends on the values of the sliders.

EXPLORING FAMILIES OF PARABOLAS

Q2 If a is positive, the parabola opens upward; if a is negative, the parabola opens downward. The larger

the absolute value of a, the narrower the parabola. The closer a is to zero, the wider the parabola.

Q3 When you drag a, the two x-intercepts remain fixed. One x-intercept exactly matches the value of slider r_1, while the other exactly matches the value of slider r_2.

Q4 Dragging either r slider changes the x-intercept corresponding to that slider.

Q5 A parabola with two equal roots has its vertex on the x-axis. Such roots are called *double roots*.

Q6 a. $y = 0.04(x + 4)(x − 6)$

b. $y = 2(x + 5)(x − 1)$

c. $y = 0.3(x − 0)(x + 3)$ or $y = 0.3x(x + 3)$

d. $y = 1(x − 3)(x + 1)$ or $y = (x − 3)(x + 1)$

e. $y = 2(x + 4)(x − 1)$

It is very important that students find the equations of these parabolas using paper-and-pencil calculations and use Sketchpad to check their answers.

Q7 The ball will land 120 feet away. The equation is $y = \left(\frac{-1}{90}\right)(x − 0)(x − 120)$ or $y = \left(\frac{x}{90}\right)(x − 120)$.

EXPLORE MORE

Q8 Because of the symmetry parabolas exhibit, the x-coordinate of the vertex is the average of the two roots, or $(r_1 + r_2)/2$. To find the y-coordinate of the vertex, substitute $(r_1 + r_2)/2$ for x in the original equation:

$$y = a\left|\frac{r_1 + r_2}{2} − r_1\right|\left|\frac{r_1 + r_2}{2} − r_2\right|$$

Simplifying, you get:

$$y = \frac{-a}{4}(r_1 − r_2)^2$$

Q9 The vertex will trace out a parabola whose vertex is at $(r_2, 0)$, opens in the opposite direction of the given parabola, and is the same shape. The equation is $y = −a(x − r_2)(x − r_2)$ or $y = −a(x − r_2)^2$.

WHOLE-CLASS PRESENTATION

Use **Factored Form Present.gsp** to present this activity to the whole class, following the directions and using the buttons in the sketch.

Parabolas in Standard Form

The vertex form of a parabola, $y = a(x - h)^2 + k$, and the factored form, $y = a(x - r_1)(x - r_2)$, both provide useful information about the graphs they describe. Perhaps the most common equation form for parabolas, though, is the standard form, $y = ax^2 + bx + c$. In some ways this is the simplest way to express a quadratic function, but it takes a little more effort to connect this form to the shape and position of the graph.

SKETCH AND INVESTIGATE

1. Open **Standard Form.gsp.** The sketch contains sliders labeled a, b, and c. You can change the values of the sliders by dragging the points at their tips.

First, create a function, evaluate it for an input variable, and plot the resulting point.

2. Use the slider values to define the function $f(x) = ax^2 + bx + c$. To do this, choose **Graph | New Function.** The New Function Calculator appears. If necessary, move it so that you can see the measurements of a, b, and c.

To enter a, b, and c, click their measurements in the sketch. To enter x, click the x key in the dialog box.

3. Enter $a*x^2 + b*x + c$ and click OK. Sketchpad creates the function.

4. Select P and measure its x-coordinate by choosing **Measure | Abscissa (x).**

5. Calculate $f(x_p)$, the value of function f evaluated at x_p. To do so, choose **Measure | Calculate.** Click the function object $f(x)$ and then measurement x_p. Finally, click OK. The value of the function appears.

6. To plot the point, select x_p and $f(x_p)$ and choose **Graph | Plot As (x, y).**

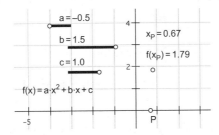

To turn tracing on or off, select the plotted point and choose **Display | Trace Plotted Point.**

Q1 Drag P back and forth along the x-axis to change the input variable for the function. How does the plotted point behave? Turn on tracing for the plotted point to better observe its behavior. What shape does this function trace out?

Traces are temporary in Sketchpad. Instead of using traces, you'll now make the graph permanent so you can compare the graphs of different functions.

7. Turn off tracing for the plotted point. Then choose **Display | Erase Traces** and drag P to make sure the traces no longer appear.

8. Plot a permanent graph of the function by selecting the function and choosing **Graph | Plot Function.** Drag P back and forth to make sure that the graph really corresponds to the path of the plotted point.

EXPLORING FAMILIES OF PARABOLAS

To make a line dashed, select it and choose **Display | Line Width | Dashed.**

By dragging point P, you explored how x and $f(x)$ vary for one particular function with specific values of a, b, and c. Now you'll change the values of a, b, and c, changing the function itself and allowing you to explore whole families of parabolas.

Q2 Adjust slider a and describe its effect on the parabola. Discuss the effect of a's sign (whether it's positive or negative), its magnitude (how big or small it is), and anything else that seems important.

Q3 Dragging a appears to change all the points on the parabola but one: the y-intercept. Adjust each slider in turn and observe the effect on the y-intercept. How is the location of the y-intercept related to the values of the three sliders?

Q4 Adjust slider c. Describe how the parabola is transformed as c changes.

Q5 Adjust slider b. What happens to the parabola as b changes?

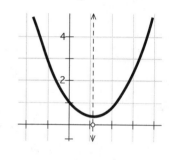

The *axis of symmetry* is the line that you can flip a parabola across without changing its position. In the next several steps you'll show that the axis of symmetry for any parabola in standard form is the vertical line that passes through the point $(-b/(2a), 0)$ on the x-axis.

To enter a and b, click on their measurements in the sketch.

9. For the x-coordinate of this point, use the Calculator to compute $-b/(2a)$.

10. For the y-coordinate, choose **Graph | New Parameter.** Name the new parameter *zero* and set its value to 0.

11. Select in order the $-b/(2a)$ calculation and the parameter *zero*, and choose **Graph | Plot As (x, y).** The axis of symmetry should pass through this point.

12. To actually construct the axis, select both the plotted point and the x-axis, and choose **Construct | Perpendicular Line.** Label this line *axis.*

13. Make line *axis* dashed, and mark it as a mirror by selecting it and choosing **Transform | Mark Mirror.**

To verify that this line really is the axis of symmetry, you can put a point on the graph and reflect the point across the line.

14. Use the **Point** tool to construct a point on the graph. Label it *Q*. Then with the point selected choose **Transform | Reflect.**

Q6 Drag point *Q* along the graph. What do you notice about the reflected point?

Q7 Change the shape of the parabola by dragging sliders *a*, *b*, and *c*. What happens to line *axis* as you change the parabola's shape? Drag point *Q* again. How does the plotted point behave?

Q8 Write the equations in standard form for three parabolas with different *a* values that have the same axis of symmetry. Check your work by adjusting the sliders in the sketch.

You can do some of these problems more easily by starting with the vertex form or factored form and converting to standard form when you're done.

Q9 Write the equation in standard form $ax^2 + bx + c$ for each of the parabolas described. As a check, adjust the sliders so that the parabola is drawn on the screen.

  a. vertex at $(2, 2)$; *y*-intercept at $(0, 4)$

  b. vertex at $(-2, 3)$; contains the point $(0, 11)$

  c. *x*-intercepts at $(-3, 0)$ and $(5, 0)$; vertex at $(1, -4)$

  d. contains the points $(0, -4)$, $(1, -1)$, and $(2, 1)$

EXPLORE MORE

Q10 The quadratic formula, which gives the roots of a parabola in terms of *a*, *b*, and *c*, is usually written

$$x_{\text{roots}} = \frac{-b \pm \sqrt{b^2 - 4ac}}{2a}$$

By the rules of fraction addition, the right side can also be written as

$$\frac{-b}{2a} \pm \frac{\sqrt{b^2 - 4ac}}{2a}$$

You've already seen that the first part of this expression represents the *x*-value of the axis of symmetry. Show that the second part of the expression represents the distance from the axis of symmetry to each of the roots.

Parabolas in Standard Form

Objective: Students plot the graph of a quadratic function in standard form, investigate the relationship between this form and its graph, and use their observations to create functions from various descriptions of their graphs.

Student Audience: Algebra 1/Algebra 2

Prerequisites: Students should have some experience with graphing quadratic equations.

Sketchpad Level: Intermediate. Students create and evaluate a function, plot and trace a point, and plot a function. They also create calculations and reflect a point across a mirror.

Activity Time: 35–45 minutes. The activity will take less time if students have already done an activity on the vertex form or the factored form. You can skip Q9 to save time.

Setting: Paired/Individual Activity (use **Standard Form.gsp**) or Whole-Class Presentation (use **Standard Form Present.gsp**)

Related Activities: Parabolas in Factored Form, Parabolas in Vertex Form, Changing Quadratic Function Forms

It's important that students understand that a parabola remains a parabola regardless of the form of its equation. Any parabola's equation can be written in many different forms. The particular forms we study are the ones that are especially simple or that convey special information.

SKETCH AND INVESTIGATE

Q1 As you drag *P*, the plotted point moves up or down, staying directly above or below *P*. This makes sense, because the point was constructed to have the same *x*-value as *P*. The plotted point follows the shape of a parabola. The precise shape depends on how the student has set the three parameters.

EXPLORING FAMILIES OF PARABOLAS

Q2 If *a* is positive, the parabola opens up; if *a* is negative, the parabola opens down. The larger the absolute value of *a*, the narrower the parabola. The closer *a* is to zero, the wider the parabola.

Q3 Neither parameter *a* nor parameter *b* has any effect on the *y*-intercept. The value of parameter *c* exactly matches the *y*-intercept.

Q4 Adjusting slider *c* moves the parabola up and down. The transformation is a vertical translation.

Q5 This is a translation. The shape and orientation of the parabola stay constant, but the axis of symmetry and both the *x*- and the *y*-coordinates of the vertex change. The path of the vertex is itself a parabola.

Q6 The reflection of point *Q* also remains on the graph, but on the opposite side.

Q7 As the parabola changes shape, the axis moves so that it always goes through the vertex, splitting the parabola into two halves that are mirror images. Dragging *Q* confirms that its reflection stays on the parabola no matter what the shape of the parabola is.

Q8 There are many possible answers. For example, each of the following has an axis of symmetry of $x = 2$: $y = 1x^2 - 4x + 13$, $y = -1x^2 + 4x - 10$, and $y = 3x^2 - 12x + 127$. (The value of *c* is irrelevant in every case—it doesn't affect the axis of symmetry.)

Q9 a. $y = 0.5x^2 - 2x + 4$

   b. $y = 2x^2 + 8x + 11$

   c. $y = 0.25x^2 - 0.5x - 3.75$

   d. $y = -0.5x^2 + 3.5x - 4$

Make sure that students find the equations of these parabolas using paper-and-pencil calculations and use Sketchpad to check their answers.

EXPLORE MORE

Q10 One approach is to make measurements in the sketch and to verify the statement numerically. For instance, you can use the Calculator to calculate the *x*-coordinate of one of the *x*-intercepts using the quadratic formula. Use **Plot As (x, y)** to plot the *x*-intercept, using the zero parameter for its *y*-coordinate. Now measure the distance from the *x*-intercept to the axis of symmetry using **Measure | Coordinate Distance.** Compare the result with the result you get when calculating the value of the second part of the quadratic formula.

To derive the result algebraically, it's easier to use the factored form of the equation. See the activity Changing Quadratic Function Forms for some ideas on how to proceed.

Changing Quadratic Function Forms

Below are three widely used forms of a quadratic function.

Standard: $ax^2 + bx + c$ Vertex: $a(x - h)^2 + k$ Factored: $a(x - r_1)(x - r_2)$

Which is better? That really depends on the situation. What information do you have? What do you need? It can even depend on your own personal style. Whatever the case, it is useful to be able to switch from one form to the other.

VERTEX TO STANDARD

Converting a function from vertex form to standard form is really just a matter of expanding the vertex expression.

$$ax^2 + bx + c = a(x - h)^2 + k$$

$$ax^2 + bx + c = a(x^2 - 2hx + h^2) + k$$

$$ax^2 + bx + c = ax^2 - 2ahx + ah^2 + k$$

If both sides express the same function, corresponding coefficients must be equal.

$$a = a \qquad b = -2ah \qquad c = ah^2 + k$$

As you may already know, the parameter *a* has the same effect on all three quadratic function forms. Perhaps that's why we use the same variable name in all three versions.

1. Open **Changing Forms.gsp.** The Vertex To Standard page already has a quadratic function in vertex form along with its graph. Change the parameters to verify that everything works as you would expect.

To use the Calculator, choose **Measure | Calculate.**

2. Use Sketchpad's Calculator to compute $-2ah$, and change the label of the result to *b*. Use the Calculator again to compute $ah^2 + k$, and change this label to *c*.

You will not be able to change the values of *b* or *c* directly. The parameters *a*, *h*, and *k* are still controlling everything.

3. Press the *Hide Vertex Form Graph* button. Choose **Graph | Plot New Function.** Enter the function in standard form: $g(x) = ax^2 + bx + c$.

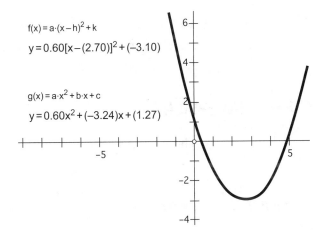

$f(x) = a \cdot (x - h)^2 + k$

$y = 0.60[x - (2.70)]^2 + (-3.10)$

$g(x) = a \cdot x^2 + b \cdot x + c$

$y = 0.60x^2 + (-3.24)x + (1.27)$

By using the Hide/Show button, you can compare the two graphs to verify that the two function forms are equivalent. If the graphs do not overlay precisely, go back and check your calculations.

4. In this sketch you can find the red text "$=y = \{1\}x^2 + (\{2\})x + (\{3\})$". This is a template, which will help complete the presentation. Select in order the template and the calculations a, b, and c. Choose **Edit | Merge Text.**

Q1 What is $2(x - 8)^2 + 5$ in standard form?

STANDARD TO VERTEX

Converting standard form to vertex form is a bit more involved. Naturally, a is still a. One way to find b and c is to make use of these two equations from the previous section:

$$b = -2ah \qquad c = ah^2 + k$$

Use the first equation to write h in terms of a and b. Substitute the result into the second equation and write k in terms of a, b, and c.

Q2 What are h and k in terms of a, b, and c?

5. Open the Standard To Vertex page. Using the techniques from the previous section, construct a presentation for converting standard form to vertex form. Here is a summary of the steps:

• Create calculations for h and k.

• Create and plot a new function in vertex form using a, h, and k.

• Using the template and the calculations, merge the text and present the new function form.

Q3 Change parameter a to zero. The vertex form graph should disappear entirely. Does this mean that you can express certain quadratic functions in standard form but not vertex form? Explain.

OTHER CONVERSIONS

There are still four function conversions left. Open the document pages in order and use the summary in step 5 as a guide.

Factored to Standard

Moving from factored form to standard form works the same way as the conversion you did in the first section. Expand the expression.

Q4 How did you define b and c?

Standard to Factored

Remember that in factored form, $a(x - r_1)(x - r_2)$, the parameters r_1 and r_2 are the roots of the function. Use the quadratic formula to find the roots in terms of a, b, and c.

Q5 How did you define r_1 and r_2?

Q6 Under certain conditions, you cannot convert the function from standard form to factored form. What are those conditions? Explain.

Vertex to Factored

As in the previous case, the parameters r_1 and r_2 are the roots. You can find the roots by setting the function equal to zero and solving for x. Find the two solutions to this equation:

$$a(x - h)^2 + k = 0$$

Q7 How did you define r_1 and r_2?

Q8 Under what conditions is it not possible to convert a vertex form quadratic function into factored form?

Factored to Vertex

There is a geometric shortcut for this conversion. The coordinates of the vertex are (h, k) and the x-intercepts are r_1 and r_2. The vertex is on the axis of symmetry of the parabola. The x-intercept points are reflections of each other across the axis of symmetry. Therefore, the x-coordinate of the vertex must be the mean of the roots. That gives you h. The parameter k is $f(h)$.

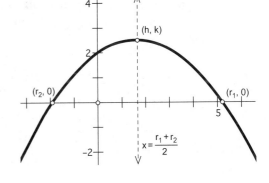

Q9 How did you define h and k?

Changing Quadratic Function Forms

Objective: Students rewrite functions in order to change them between standard, vertex, and factored forms. They edit sketches so that the changes happen interactively.

Student Audience: Algebra 2/Precalculus

Prerequisites: This activity is quite challenging in terms of both algebra skills and Sketchpad skills. Students should be familiar with the properties of all three forms of quadratic functions, and they must be able to perform symbolic manipulations of algebraic expressions.

Sketchpad Level: Challenging. Students perform a number of calculations on the screen. They also plot functions and use prepared templates to merge text.

Activity Time: 50+ minutes. The activity is broken into six similar tasks. It has extensive guidance on the first, and guiding hints after that. The end of each task is a clean breaking point, so you could abbreviate the activity or perform it over more than one session.

Setting: Paired/Individual Activity (use **Changing Forms.gsp**)

A few students may get through this quickly, but most will need a lot of time. It will be especially difficult for students who have poor computer skills. With so many expressions to enter, they may tend to make mistakes with order of operations. It is important not to rush them. Encourage them to work together.

VERTEX TO STANDARD

Q1 $2x^2 - 32x + 133$

STANDARD TO VERTEX

Q2 $h = -\dfrac{b}{2a}$ $k = \dfrac{4ac - b^2}{4a}$

Q3 When $a = 0$, the formulas for both h and k are undefined. However, it is also true that when $a = 0$, the standard form function is linear, not quadratic. In fact, you can rewrite any quadratic function from standard form to vertex form.

OTHER CONVERSIONS

Factored to Standard

Q4 $b = -a(r_1 + r_2)$ $c = ar_1r_2$

Standard to Factored

Q5 $r_1 = \dfrac{-b - \sqrt{b^2 - 4ac}}{2a}$ $r_2 = \dfrac{-b + \sqrt{b^2 - 4ac}}{2a}$

Q6 The condition $a \neq 0$ occurs here once more, but again, that would only exclude functions that are not quadratic anyway. The only real problem is having a negative discriminant. You cannot use factored form if $b^2 - 4ac < 0$.

Vertex to Factored

Q7 $r_1 = h - \sqrt{-\dfrac{k}{a}}$ $r_2 = h + \sqrt{-\dfrac{k}{a}}$

Q8 In this case, either k and a must have opposite signs, or k must be zero. Otherwise, r_1 and r_2 would both be undefined. This happens only when the function has no real roots.

Factored to Vertex

Q9 $h = \dfrac{r_1 + r_2}{2}$ $k = -\dfrac{a(r_1 - r_2)^2}{4}$

Students may give the formula for k in various unsimplified forms. They may even answer simply $k = f(h)$. If they have calculated h, then it is possible to do that in Sketchpad.

The Discriminant

All quadratic functions have zero, one, or two real roots. The discriminant is a useful calculation that tells you how many roots there are for a given quadratic function.

This is the general definition of a quadratic function in standard form:

$$f(x) = ax^2 + bx + c, \text{ where } a \neq 0$$

If the function has roots, you can find them with these formulas:

$$r_1 = \frac{-b - \sqrt{b^2 - 4ac}}{2a} \qquad r_2 = \frac{-b + \sqrt{b^2 - 4ac}}{2a}$$

The discriminant is the part under the radical signs. We often write it as Δ (delta).

$$\Delta = b^2 - 4ac$$

SKETCH AND INVESTIGATE

1. Open **Discriminant.gsp.** It has the graph of the quadratic function $f(x)$. Above the function definition are parameters a, b, and c.

2. Experiment by changing the parameters and observing the changes in the graph. When you finish, set the parameters so that the function now has this definition:

$$f(x) = x^2 - 2x - 3$$

To show the Calculator, choose **Measure | Calculate.**

3. Use Sketchpad's Calculator to compute the two roots and the discriminant, using the formulas above and the parameters on the screen.

To put a subscript at the end of a label, enclose the subscript in square brackets. For instance, r[1] will appear as r_1.

4. Label the calculations r_1, r_2, and *delta.*

Q1 What are the roots and the discriminant? Does the graph confirm the root calculations? Explain.

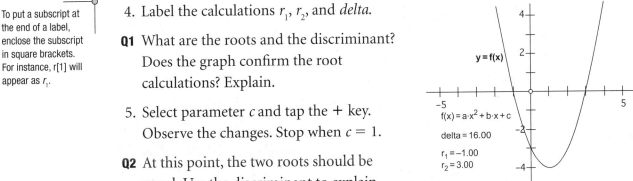

5. Select parameter c and tap the + key. Observe the changes. Stop when $c = 1$.

Q2 At this point, the two roots should be equal. Use the discriminant to explain why this is true. Describe the graph.

6. Increase parameter c further. Let $c = 2$.

Q3 What are the values of the discriminant and the roots? How does the discriminant explain the roots? Describe the graph.

GRAPH INTERSECTIONS

It is often possible to use the discriminant to find information about the intersections of two curves. The idea is to first work the equations into a single quadratic equation. Consider the graphs of these two general functions, one parabola and one line:

$$f(x) = a_1x^2 + b_1x + c_1 \qquad g(x) = mx + u$$

At any point where these graphs intersect, their y-coordinates must be equal. Therefore, you can set the functions equal to each other:

$$a_1x^2 + b_1x + c_1 = mx + u$$

Q4 Group the terms of this equation to form a quadratic function on the left side and zero on the right:

$$(\ )x^2 + (\ )x + (\ ) = 0$$

7. Go to page 2. Functions $f(x)$ and $g(x)$ are defined as shown above. Based on your answers to Q4, use the Calculator to compute values for a, b, and c.

8. Use the parameters from the previous step to calculate the discriminant.

Q5 Change the parameters of functions $f(x)$ and $g(x)$, and observe the graph. What do you see when the discriminant is positive? Zero? Negative?

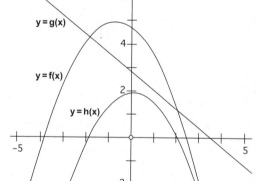

9. Using the parameters from step 7, plot another quadratic function:

$$h(x) = ax^2 + bx + c$$

Q6 What is the relationship between the roots of function $h(x)$ and the intersection points of $f(x)$ and $g(x)$?

TANGENCY

10. Hide the graph of $h(x)$. Change the parameters so that function $f(x)$ has the definition below:

$$f(x) = -2x^2 - 3x + 2$$

Q7 What is the y-intercept of the graph of $f(x)$? Now change the parameters of $g(x)$ so that the two graphs have the same y-intercept and are tangent at that point. What is the new definition of $g(x)$? Explain how you got it.

Exploring Algebra 2 with The Geometer's Sketchpad
© 2007 Key Curriculum Press

The Discriminant

Objective: Students calculate the discriminant of a quadratic function and observe its relationships to the roots of the function and to curve intersections and tangency.

Student Audience: Algebra 2/Precalculus

Prerequisites: Students should be familiar with quadratic functions, their graphs, and the quadratic formula. Previous work with the discriminant is not necessary.

Sketchpad Level: Intermediate. Most of the work involves calculations and graphing.

Activity Time: 30–40 minutes

Setting: Paired/Individual Activity (use **Discriminant.gsp**) or Whole-Class Presentation (use **Discriminant Present.gsp**)

SKETCH AND INVESTIGATE

3. Students can simplify these calculations by finding delta first and then using that in the calculations for r_1 and r_2.

Q1 The roots are -1 and 3. The discriminant is 16. Since the roots are the solutions to the equation $f(x) = 0$, these are also the x-intercepts of the graph.

Q2 The only difference between the two root formulas is the sign of the radical. If the discriminant is zero, then the radical effectively goes away.

$$r_1 = \frac{-b - \sqrt{b^2 - 4ac}}{2a} = \frac{-b - \sqrt{0}}{2a} = -\frac{b}{2a}$$

$$r_2 = \frac{-b + \sqrt{b^2 - 4ac}}{2a} = \frac{-b + \sqrt{0}}{2a} = -\frac{b}{2a}$$

The graph is tangent to the x-axis at $(1, 0)$.

Q3 The discriminant is -4. Both of the roots are undefined. When the discriminant is negative, the root formulas include the square root of a negative number, which is why they are undefined. Since the function has no roots, its graph has no x-intercepts.

GRAPH INTERSECTIONS

Q4 $a_1 x^2 + (b_1 - m)x + (c_1 - u) = 0$

7. Check in with students at this step. If they do not correctly define the parameters of the new function, nothing else in the activity will work out right.

$$a = a$$
$$b = b_1 - m$$
$$c = c_1 - u$$

Q5 When the discriminant is positive, the graphs intersect in exactly two points. When the discriminant is zero, the graphs are tangent. When the discriminant is negative, the graphs do not intersect.

Q6 The roots of $h(x)$ are the x-coordinates of the intersection points of the graphs of $f(x)$ and $g(x)$. Graphically, this means the intersection points are horizontally aligned with the x-intercepts of the graph of $h(x)$.

TANGENCY

Q7 The y-intercept of $f(x)$ is 2, the constant term of the function. In order to give $g(x)$ the same y-intercept, let $u = 2$. The graphs will be tangent if and only if the discriminant is equal to zero. Adjust the slope, m, until the discriminant is equal to zero.

$$g(x) = -3x + 2$$

Students may notice that these are the last two terms of $f(x)$. In fact, this works for any quadratic function. The last two terms define a linear function that shares the same y-intercept and is tangent at that point.

After finding that $u = 2$, students might find m analytically, by setting the discriminant equal to zero.

$$b^2 - 4ac = 0$$
$$(b_1 - m)^2 - 4a_1(c_1 - u) = 0$$
$$(-3 - m)^2 - 4(-2)(2 - 2) = 0$$
$$(-3 - m)^2 = 0$$
$$m = -3$$

The Discriminant

The purpose of this presentation is to show the connections between quadratic functions and their discriminants. The discriminant can give quite a lot of information regarding roots, intersections, and tangency.

A SIMPLE EXAMPLE

1. Open **Discriminant Present.gsp.** Page 1 starts with a quadratic function written in standard form. Briefly review the definition.

2. Press the *Show Discriminant* button to show that definition, and press *Show Roots* to show where the discriminant appears in the quadratic formulas.

Q1 What happens if the discriminant is zero? (The roots are equal.) What happens if the discriminant is negative? (Both roots are undefined.)

3. Go to page 2. This is the graph of a quadratic function. The parameters *a, b,* and *c* control the coefficients of the function definition. Select parameter *c* and tap the + key.

Q2 As the graph moves upward, stop occasionally to ask the same questions. How many roots are there? Why? In each case, show how the number of roots is related to the sign of the discriminant. Give special attention to the tangency that occurs when the discriminant is zero.

4. Go to page 3. This shows two functions, one quadratic and one linear. Lead the class into a discussion of the intersection points of their graphs. For any given *x,* the functions represent the *y*-coordinates of the corresponding points on their graphs. At any intersection point, the graphs must have the same *y*-coordinate.

Q3 Press the $f(x) = g(x)$ button to set the functions equal. What do you get when you group like terms on the left side of the equation? (A single quadratic function is equal to zero.) Press *Group Terms* to do the same thing on the screen. Press *Show h(x)* and *Show Parameters* to identify the parameters of the new function.

5. Go to page 4. This is a graphic representation of the same problem, a quadratic $f(x)$ and a linear $g(x)$. Press *Show h(x)* to show the function that was derived on the previous page. Its discriminant will also appear. Change the function parameters.

Q4 What is the relationship between the discriminant and the intersections of $f(x)$ and $g(x)$? (The graphs have two intersection points when the discriminant is positive, are tangent when it is zero, and do not intersect when it is negative.)

Q5 How do the intersections relate to the graph of $h(x)$? (They are horizontally aligned with the zeroes of $h(x)$.) Press *Show Intersections* to illustrate this fact.

Exploring Algebra 2 with The Geometer's Sketchpad
© 2007 Key Curriculum Press

Parabolas: A Geometric Approach

You may think of algebra and geometry as two very different branches of mathematics. In many ways they are. But you've seen that algebraic equations, such as $y = x + 3$, can be graphed as lines, which are geometric objects. Now you're studying parabolas—the graphs of equations such as $y = x^2 + 3$. Can parabolas be described geometrically, without using algebraic equations? In this activity you'll see that they can.

FOCUS AND DIRECTRIX

A *circle* can be described as the set of points in a plane that are the same distance from a fixed point—the center. Similarly, a *parabola* can be described as the set of points in a plane that are the same distance from a fixed point—the *focus*—as from a fixed line—the *directrix*.

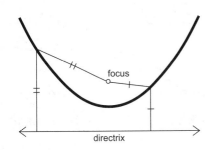

This is a more complicated definition than the circle's, but as you do this activity it should make more and more sense.

A PARABOLA FROM SCRATCH

If the **Line** tool isn't showing, press down on the current **Straightedge** tool and choose the **Line** tool from the palette that pops up.

1. In a new sketch choose **Edit | Preferences.** On the Text panel, check Show Labels Automatically For All New Points. On the Color panel, check Fade Traces Over Time.

2. Using the **Line** tool, construct a horizontal line *AB*. (To make it exactly horizontal, hold the Shift key while you construct.) The line should be about a third of the way from the bottom of the sketch window. Show the line's label and change it to *directrix*.

3. Use the **Segment** tool to construct segment *CD* where *C* is on the line and *D* is about an inch above it.

Click on an object with the **Text** tool to show its label. Double-click the label itself to change it.

4. Change point *D*'s label to *D: Focus.*

5. Construct the midpoint of segment *CD* by selecting it and choosing **Construct | Midpoint.**

6. Select segment *CD* and midpoint *E*, and choose **Construct | Perpendicular Line.**

After step 6

The line you just constructed—perpendicular to segment *CD* through its midpoint—is called the *perpendicular bisector* of segment *CD*.

Q1 Imagine a point anywhere on the perpendicular bisector. How do you think the point's distance to *C* compares to its distance to *D*?

7. Construct a line perpendicular to $\overleftrightarrow{AB}$ and passing through point *C*.

8. Select the two perpendicular lines (the ones constructed in the last two steps) and choose **Construct | Intersection.**

You just constructed point *F*—the first point on the parabola.

To hide objects, select them and choose **Display | Hide.**

9. Hide $\overleftrightarrow{FC}$. Then, using the **Segment** tool, construct $\overline{FC}$ and $\overline{FD}$.

10. Measure the lengths of the two new segments. To do this, select them and choose **Measure | Length.**

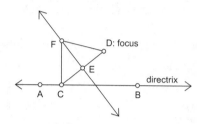

After step 9

Q2 Using the **Arrow** tool, drag *C* back and forth along line *AB*.

Refer back to the parabola definition in the introduction.

What do you notice about the length measurements as you drag *C*? Explain how this demonstrates that *F* is indeed always a point on the parabola defined by the given focus and directrix.

11. Select point *F* and choose **Display | Trace Intersection.** Now once again drag *C* back and forth. Point *F* will leave a trail as it traces out the parabola.

12. Drag the focus away from the directrix. Again, drag point *C*. Notice how this curve compares to the previous one. Now drag the focus closer to the directrix than it was originally and repeat the process.

Tracing the curve in this way works well enough to show you the shape of the parabola. But it can get a little annoying having to drag point *C* again and again. Here's a more efficient approach.

Select the point and choose **Display | Trace Intersection** *to turn tracing on or off.*

13. Turn off tracing for point *F*.

14. Select points *F* and *C*, and choose **Construct | Locus.**

The entire curve appears. It's called the *locus* of point *F* as point *C* moves along $\overleftrightarrow{AB}$.

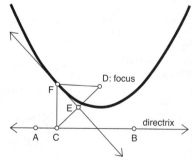

To hide objects, select them and choose **Display | Hide.**

15. To make the diagram less cluttered, hide everything except the parabola, the focus, and the directrix.

16. Drag the focus around and observe how the curve changes.

Q3 What happens to the parabola as the focus is dragged farther away from the directrix? Closer to it?

Q4 What happens to the parabola when the focus is dragged below the directrix?

17. Save your sketch. You may want to use it for another activity, Building Headlights and Satellite Dishes.

EXPLORE MORE

Q5 Use the commands in the Construct menu to construct the vertex of your parabola. This point should remain the vertex no matter where you drag the focus and the directrix.

Q6 Show that the equation of a parabola with its vertex at the origin and its focus at $(0, p)$ is $y = \frac{x^2}{4p}$. To do this, select the vertex you constructed in Q5 and choose **Define Origin** from the Graph menu. Then measure the y-coordinate (ordinate) of the focus and relabel this measurement p. Now use **Graph | Plot New Function** to plot $\frac{x^2}{4p}$. This plot should be right on top of the parabola, even if you drag the focus or the directrix.

Parabolas: A Geometric Approach

Objective: Students use a locus definition to construct a parabola geometrically and relate the result to parabolas that are plotted algebraically.

Student Audience: Algebra 1/Algebra 2

Prerequisites: This activity works best if students have already been learning about parabolas from an algebraic perspective, but it also works as a first introduction to parabolas.

Sketchpad Level: Intermediate

Activity Time: 30–45 minutes depending on students' Sketchpad proficiency

Setting: Paired/Individual Activity (The document **Parabolas Geometric.gsp** shows the construction at several stages.)

Related Sketch: Conic Sections.gsp allows you to vary the angle at which a plane intersects a cone and to view the intersection from various directions.

A good way to introduce this activity might be to talk a little about the history of parabolas. Parabolas have been studied for over 2000 years, but algebraic equations for them have only been known about for the past 500 years or so. How did people describe and define parabolas before? How were they drawn without coordinate geometry? Where do parabolas come from? (They are sections of cones created by a cutting plane parallel to the side of a cone). This kind of historical perspective may help students appreciate the true power of the algebraic equations.

CONSTRUCTION TIPS

2. Holding down the Shift key while drawing the line helps keep it horizontal.

3. Perform your first click on the directrix itself (though not on one of its control points). (The directrix should be highlighted before you click.)

5. Use the **Arrow** tool in this step.

7. Refer to step 6 for a reminder how to do this.

11. Make sure only point *F* is selected here. To deselect all objects, click in blank space.

12. Traces will gradually fade because of the Preference setting made in step 1. If you prefer that traces not fade, uncheck Fade Traces Over Time on the Color panel of the Preferences dialog box. To clear traces from the screen, choose **Display | Erase Traces.**

13. Again, make sure only *F* is selected.

A PARABOLA FROM SCRATCH

Q1 It's the same distance from either endpoint. This is a theorem from geometry and can be proven, but for our purposes it's probably okay just to leave this up to students' common sense.

Q2 The two length measurements are always equal to each other. (This should confirm the students' conjectures from Q1.) The length of segment *FC* is the distance from *F* to the directrix, and the length of *FD* is the distance from *F* to the focus. Point *F*, then, fits the definition from the activity's introduction and is therefore on the parabola defined by the given focus and directrix.

Q3 The farther point *C* is from the directrix, the wider the parabola; the closer, the skinnier.

Q4 When point *C* is below the directrix, the parabola opens downward.

EXPLORE MORE

Q5 Select the focus and the directrix, and choose **Construct | Perpendicular Line.** Click on the spot where this new line intersects the directrix to construct the point of intersection there. Hide the line and then construct a segment between the focus and the new point. Construct the midpoint of this segment using **Construct | Midpoint.** This point is the vertex of the parabola.

Parabolas in Headlights and Satellite Dishes

Have you ever wondered why satellite dishes are shaped the way they are? And what is it about the way headlights are designed that makes the light travel outward in one direction? It turns out that both devices use *parabolic reflectors* because of their special reflective properties. In this activity you'll construct a two-dimensional model of a parabolic reflector and explore what makes it ideal for reflecting and collecting rays.

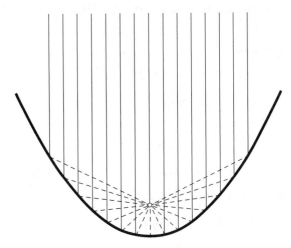

BACKGROUND

The two angles are measured relative to the *normal* line—the line that's perpendicular to the tangent.

In order for this activity to make sense, there are two things you need to know about the *Law of Reflection*. First, when a ray of light reflects from a flat surface, such as a mirror, the *angle of incidence* (∠1 below) equals the *angle of reflection* (∠2 below). This is a fancy-sounding law of physics that states something very simple: Light bounces off at the same angle it hits. Second, if light bounces off a *curved* surface, the Law of Reflection still holds—just imagine the light bouncing off a line *tangent* to the curve.

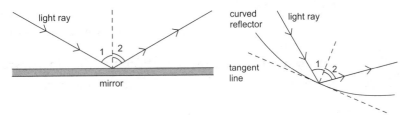

SKETCH AND INVESTIGATE

1. If you have completed the activity Parabolas: A Geometric Approach, open your sketch from that activity. Otherwise, open **Reflector.gsp** and skip to step 4.

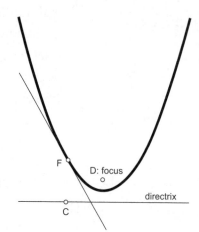

2. Choose **Display | Show All Hidden.** Then hide all but the following objects: the parabola, focus, directrix, points *C* and *F*, and the line that's tangent to the parabola.

To hide objects, select them and choose **Display | Hide.**

3. To make the parabola flatter, drag the directrix down near the bottom of the window and the focus about two-thirds of the way to the top.

You need a light source—the tiny bulb inside a headlight whose light bounces off the reflector into the outside world.

4. Using the **Point** tool, draw a new point *G* inside the parabola somewhere not too far from the focus.

Now you need a ray of light coming from the source and hitting the reflector.

Either use the **Segment** tool or select the two points and choose **Construct | Segment.**

5. Construct a segment between points *G* and *F*.

The bulb actually sends out rays of light in all directions. Use a Sketchpad locus construction to show 15 of these rays.

6. Select the new segment and point *C*, and choose **Construct | Locus.** With the locus selected, choose **Display | Line Width | Dashed.** With the locus still selected, choose **Edit | Properties.** On the Plot panel, set the number of samples to 15.

You'll see 15 rays coming from the light source and hitting the reflector.

The normal line is perpendicular to the tangent at the point where the ray hits the reflector.

Each ray must bounce off the reflector at the same angle at which it hit the reflector. The easiest way to construct this involves the *normal* line.

7. To construct the normal line, select point *F* and the line that's tangent to the parabola. Then choose **Construct | Perpendicular Line.**

The symmetry of reflection guarantees that the incoming ray and the outgoing ray will both make the same angle with the normal.

8. Double-click the new line to mark it as a mirror for reflection. Now select point *G* and choose **Transform | Reflect.** The reflected image of point *G*, *G'*, appears.

9. Select in order point *F* and point *G'*. Then choose **Construct | Ray.**

This ray is the original ray of light after it's bounced off the reflector. By reflecting *G* across the perpendicular line, you guaranteed that the angle of incidence would equal the angle of reflection, as the Law of Reflection states.

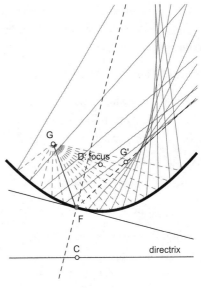

After step 10

10. Repeat step 6 using the new ray in place of the segment. Give the new locus a thin line weight (instead of dashed) to distinguish it from the other, and set the number of samples to 15 as well.

You should now see the 15 rays of light coming from the light source and bouncing off the reflector.

Q1 Drag the light source around and observe how its location affects the reflected light. Where does the light source need to be in order for the light to travel out in parallel rays, as with car headlights?

11. Merge the light source and the focus. To do this, select these two points and choose **Edit | Merge Points.**

The two points merge into one. Now the light source is at the focus and light is bouncing off in parallel rays.

Q2 Drag the focus and the directrix separately to change the shape of the reflector. Do the reflected rays remain parallel?

12. Clean up the sketch by hiding segment *GF*, the perpendicular line through point *F*, the ray originating at *F*, point *G'* defining that ray, and anything else you'd like out of the way.

Q3 A satellite dish operates similarly to a car's headlights, except that instead of emitting rays, it collects them. Explain how you would design a satellite dish.

EXPLORE MORE

Q4 Do some research to find other objects shaped like a parabolic dish. Explain why each of them uses this shape.

Parabolas in Headlights and Satellite Dishes

Activity Notes

Objective: Students construct a two-dimensional model of a parabolic reflector and explore what makes it ideal for reflecting and collecting rays.

Student Audience: Algebra 1/Algebra 2

Prerequisites: This activity can stand on its own, or you can use it as a follow-up to the activity Parabolas: A Geometric Approach.

Sketchpad Level: Challenging. There are quite a few construction steps.

Activity Time: 40–50 minutes

Setting: Paired/Individual Activity (use **Reflector.gsp** if students don't have sketches from the activity Parabolas: A Geometric Approach)

Related Activities: Parabolas: A Geometric Approach, Conic Reflections

One way to introduce this activity is to ask students what shape a "burning mirror"—a mirror designed to reflect the sun's rays to one point—should be. Why would a straight line (flat mirror) be a bad choice? Students may be interested to know that Diocles discovered—and used—the answer to this question c. 200 B.C.E. Diocles was improving on work done a century earlier by Archimedes, who, as legend has it, used burning mirrors to set the Roman fleet ablaze as it entered Syracuse harbor. Diocles' principle is still used today in solar-powered steam generators. (See Toomer, G.J., *Diocles on Burning Mirrors.* New York: Springer-Verlag, 1976. The Archimedes legend was the subject of Mythbusters Episode #46 on the Discovery Channel.)

SKETCH AND INVESTIGATE

Q1 When the light source is at the focus, the rays will reflect parallel to each other.

Q2 Provided the light source is at the focus, the rays will reflect parallel to each other (and to the axis of symmetry of the parabola) no matter how narrow or wide the parabola is.

Q3 Place the sensor at the focus of the three-dimensional parabola (called a *parabaloid*). The signals coming in roughly parallel to the direction the dish is facing will reflect right to the sensor whereas all other signals will bounce and miss the sensor.

EXPLORE MORE

Q4 Another example of a parabolic dish is the parabolic microphones often seen on the sidelines of football games and used to catch the sounds of the game. These work the same way a satellite dish works: Sound waves coming in parallel to the direction the device is aimed bounce toward the microphone, which is located at the focus. Sound waves from other directions bounce harmlessly away.

Conic Reflections

You may already know that light rays from the focus of a parabola reflect to form parallel rays. (For instance, the reflector in a searchlight or a car's headlight is shaped like a parabola.) A parabola is the graph of a quadratic function, but it is also a conic section. As it turns out, all of the conic sections have special reflective properties.

REFLECTIONS IN AN ELLIPSE

Light and sound are very different in nature but share many common properties. You can imagine either of them as waves or rays.

Kidney stones are a very painful ailment caused by crystallization of salts inside the kidney. They are in fact small stones. A modern treatment is extracorporeal shock wave lithotripsy (ESWL). Patients are happy about the extracorporeal part, because the word means "outside the body." Surgery usually is not necessary. The idea is to create a sonic pulse outside so that the shock waves converge at the stone and break it into smaller pieces that can pass out of the kidney.

1. Open the Ellipse page of **Conic Reflections.gsp.** Drag the points to see how they affect the image. This particular ellipse is defined by the two foci (F_1 and F_2) and a free point P on the curve. Point A is attached to the ellipse, and the line through A is tangent to the ellipse.

Point B represents the source of the waves of sound or light.

2. Construct point B in the interior of the ellipse. Construct line segment BA. This represents a ray of sound traveling from B and striking the surface at point A.

The locus shows the position of segment BA for various possible positions of point A on the ellipse.

3. Select point A and segment BA. Choose **Construct | Locus.** Select the locus and choose **Edit | Properties.** On the Plot panel, change the number of samples to 15. Select the locus again and change the line width to dashed. You should now see 15 rays coming from B and striking the surface at various points.

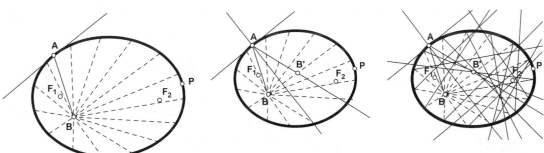

When a ray reflects from a surface, the angle of incidence is equal to the angle of reflection, and you measure those angles from a line that is perpendicular to a tangent at that point on the surface. This principle applies to all curved surfaces.

The normal line intersects the tangent, and the curve itself, at a right angle.

4. Select point A and the tangent line. Choose **Construct | Perpendicular Line.** This is called a *normal* line.

5. Double-click the normal line to mark it as a reflection line. Select point B and choose **Transform | Reflect.** Construct ray AB'. This is the reflected ray.

6. Construct the locus of the reflected ray. Again, change the number of locus samples to 15.

Q1 Point *B* is the source of the sonic pulses. Drag it around the screen and watch the pattern change. Where can you place it in order to get the reflected rays to converge at a single point? When this happens, where do the rays converge again?

Q2 You may notice that the rays come together at a point inside the ellipse. Why is this a problem for the ESWL treatment? Suggest a solution.

REFLECTIONS IN OTHER CONICS

There are still two conic sections—the circle and the hyperbola—that you have not investigated for reflective properties. The Sketchpad document contains a page for each of them. Repeat these same construction steps to explore their properties.

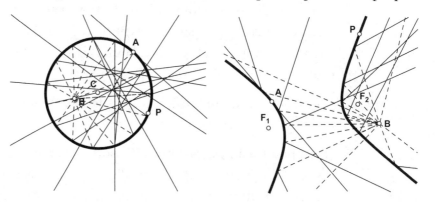

Q3 You can think of a circle as a special case of an ellipse. The two foci are in the same place, which is the circle center. With this in mind, where would you expect the ray reflections to go if their source is at the center of a circle?

Q4 The hyperbola may be a bit more difficult to see. Instead of watching where the rays go, try to see where they appear to be coming from. If the source of the rays is on one focus, where do the reflected rays appear to come from?

PRESENT

The Parabola page is already completed. It summarizes the findings from the Building Headlights and Satellite Dishes activity. On each of the other pages, use the **Text** tool to describe the special reflective properties of the curve.

Notice that the Parabola page has a button that moves point *B* to the focus. To create a similar button on the other pages, select in order point *B* and the destination point. Choose **Edit | Action Buttons | Movement.**

© 2007 Key Curriculum Press

Conic Reflections

Objective: The reflective properties of parabolas are well known. In this activity students explore interesting reflective properties of other conic sections.

Student Audience: Algebra 2/Precalculus/Geometry

Prerequisites: Students should have introductory knowledge of the names and geometric properties of conic sections. They do not need any experience with the analytical representations.

Sketchpad Level: Challenging. The activity involves several locus constructions and editing of properties.

Activity Time: 40–50 minutes

Setting: Paired/Individual Activity (use **Conic Reflections.gsp**) or Whole-Class Presentation (use **Conic Reflections Present.gsp**)

Related Activities: This activity is intended as an extension of the Parabolas in Headlights and Satellite Dishes activity. If the students have not performed that task, you should first spend some time explaining the properties shown in the Parabola page of the document. That sketch is completed, and it will give the class a better understanding of the objective.

When you review this activity with your class, ask the students to explain the connection between conic sections and kidney stones.

REFLECTIONS IN AN ELLIPSE

Q1 When you drag point *B* to one focus, the rays are concurrent at the other focus.

Q2 Remember that the rays must converge inside a person's body. If that point is inside the ellipse, then the patient must be inside. Students may guess (incorrectly) that the reflecting chamber must be big enough to hold the patient.

In practice, the chamber is nowhere near that large. It is ellipsoidal in shape, but one end is open (see below). This still allows most of the shock energy to reach the target, but some of it passes out the open end.

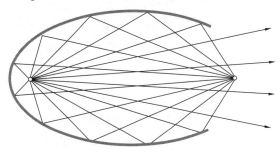

REFLECTIONS IN OTHER CONICS

Q3 A circle is an ellipse with both foci at the same center point. Therefore, rays sent from the center reflect back to the center.

Q4 In a hyperbola, send the rays from one focus. After reflection, they will travel directly away from the other focus.

Conic Reflections

To present the reflective properties of conic sections, you can use the prepared document **Conic Reflections Present.gsp.** This document allows you to skip the construction work and present a summary of the observations. In general, just walk through the pages in order, pressing the action buttons in order from the top down.

1. Open the Flat Surface page of **Conic Reflections Present.gsp.** This shows a reflection on a flat surface, like a mirror or a pool table cushion. Point *B* is the source of a ray that strikes the surface at *A*.

2. Press *Show Normal Line.* Explain the physics principle that the angle of incidence is equal to the angle of reflection, and you measure both angles with respect to the normal line.

3. Press *Show Reflection.* Identify the angles of incidence and reflection. Drag points *A* and *B* to show the changes in the reflection.

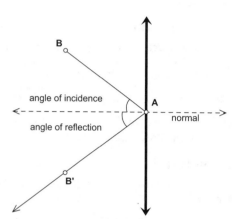

Q1 Open the Curved Surface page. What happens when we have a curved surface? How do we find the normal line? (The normal is perpendicular to the tangent at the same point *A*.)

4. Show the hidden objects. When dragging point *B*, be sure to show that the same principle holds when it is on the concave side of the curve.

Q2 The remaining pages show reflective properties of conic sections in turn. Each page opens with a single reflected ray. Press *Show Rays* to show 15 rays all originating at point *B*. No particular pattern is visible until you drag point *B* to a focus. In each case, challenge students to predict what the reflected ray pattern will be. What will happen when we drag this to the focus? What practical applications can we find for this effect?

Q3 After showing that an ellipse reflects a ray from one focus to the other, drag point *B* away from the focus and hide the rays. Press *Move F_2 to F_1* and let students see that a circle is actually an ellipse with coincident foci. With this fact in mind, what should happen when point *B* is at the center of a circle? (The reflected rays go straight back to the center.)

5. The hyperbola reflection pattern may be more difficult to see, so give students more time. When point *B* is on F_1, the reflected rays move out away from the second focus rather than through it. Press *Show Extensions* to help them see the alignment.

Modeling Projectile Motion

When you toss a ball into the air, the laws of physics control its motion through the air and back to the ground. In this activity you'll use Sketchpad's **Iterate** command to model a projectile as its position, velocity, and gravity interact.

To create the model, you'll make a sketch showing what happens to the projectile in a short period of time. You'll use iteration to repeat the construction and create a graph of the projectile's motion over a much longer period of time.

SKETCH

1. Open **Projectile Motion.gsp.** Notice the vector for gravity (g), the vector for the initial velocity (v_0) of the object, and the point for the initial position (p_0) of the object. Vector g points down to represent the downward acceleration gravity exerts on projectiles. Vector v_0 points diagonally to indicate the movement of the ball at the instant it is thrown.

At this instant, the ball has not moved from its initial position. To begin, you'll construct a segment to represent the ball's motion by using the velocity vector to find the ball's position after the first interval of time.

In the Translate dialog box, select Marked for all translations.

2. To mark the velocity vector as the vector for translation, select its initial and final points, and choose **Transform | Mark Vector.** To translate the position, select point p_0 and choose **Transform | Translate.** Label the translated position p_1 and connect the initial position and the translated position with a segment.

The new segment shows how the ball moves through the air during the first period of time. But if you've ever thrown a ball, you know that gravity affects its flight. In your sketch you must also take into account gravity's effect on the ball's velocity. To do this, you'll use the gravity vector to construct a new velocity vector.

This segment represents the velocity vector at the end of the first period of time.

3. Mark the gravity vector as the vector for translation. Translate point v_0 by this vector, and then construct a segment from the original tail of the velocity vector to v_1 (the new translated point).

The v_1 vector represents the object's velocity at the end of the first interval of time. So v_1 describes the ball's velocity when it's at p_1.

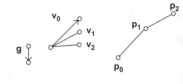

4. Repeat steps 2 and 3 for the second period of time. Be sure you construct the new position segment from the end of the old one, but the new velocity vector with its tail in the same place as the existing velocity vectors.

Q1 How did the velocity change as the ball moved from p_0 to p_1? Is the ball going faster or slower? How did the direction of its motion change?

5. Repeat steps 2 and 3 for a third period of time.

Q2 What do you notice when you compare the velocities v_0, v_1, v_2, and v_3 for the three periods? How is the velocity vector changing the object's path?

Q3 What do you notice when you compare positions p_0, p_1, p_2, and p_3? Describe the motion of the object.

Q4 Make a prediction for how the position will change during the next ten periods of time.

It would be a lot of work to do the same construction ten more times. You can use iteration to make the process easier.

Choose **Edit | Undo** repeatedly until you get back to the end of the first period.

6. Undo your work back to the end of the first period of time.

Gravity is constant, so you didn't translate that vector. This means that you only have to keep track of changes in the position and the velocity. The first step of the iteration will map $p_0 \Rightarrow p_1$ and $v_0 \Rightarrow v_1$.

Points p_0 and v_0 are called the *pre-image* points of the iteration, and points p_1 and v_1 are the *image* points.

7. To create the iteration, select the pre-image points p_0 and v_0 and choose **Transform | Iterate**. Then click p_1 and v_1 in the sketch so that each pre-image point is correctly mapped to its image point. Click Iterate to complete the iteration and show the next three steps.

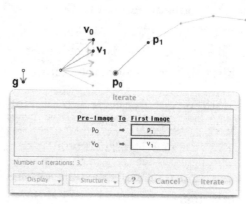

Exploring Algebra 2 with The Geometer's Sketchpad
© 2007 Key Curriculum Press

8. With the iterated images still selected, press the + key on your keyboard to increase the number of iterations until you have about 15 of them.

Q5 Describe the shape of the object's flight. For each period of time, how do the direction of the velocity vector and the path of the object relate to each other?

Q6 During which period of time does the velocity begin to point downward? Can you tell at what moment the ball begins to fall downward by looking only at the velocity vector? Can you tell when the velocity vector begins to point downward just by looking at your construction of the ball's path?

To turn gravity off and check your prediction, drag the end points of the gravity vector so that they coincide.

Q7 What do you think the path would look like if there were no gravity? Draw your prediction on paper.

Q8 How well does the shape of the path represent the actual flight of a ball through the air? Think of a time when you've seen a ball fly through the air. How does your constructed path differ from your observation of how balls fly through the air?

The model can be more accurate if you make each iteration represent a shorter period of time. To make this change, you will have to change the lengths of the original vectors.

Q9 To make each period of time half of its original value, how much shorter must you make the vectors? How many more iterations will you need to model the same period of time?

9. Shorten the gravity and initial velocity vectors and increase the number of iterations until your model shows smooth motion for the projectile.

Q10 What shape does the projectile's motion appear to have?

EXPLORE MORE

10. Using **Graph | Plot New** Function and the form $f(x) = ax^2 + bx + c$, try to match a parabola to the path of the ball. Adjust the parameters to change the shape of your parabola.

11. Extend the model to use a parameter to represent the time interval. (*Hint:* Use the parameter to dilate the original vectors before you iterate them.)

Modeling Projectile Motion

Objective: Students make a Sketchpad model of a physical process operating for a short period of time, and then iterate that process to see how it produces reasonable long-term behavior.

Student Audience: Algebra 1/Algebra 2/Physics

Prerequisites: It helps if students have seen a velocity vector used to show both how fast and in what direction an object is moving.

Sketchpad Level: Intermediate

Activity Time: 35–45 minutes

Setting: Paired/Individual Activity (use **Projectile Motion.gsp**) or Whole-Class Presentation (use **Projectile Motion Present.gsp**)

SKETCH

Q1 During the first period of time, the velocity's magnitude is reduced and its direction is less steeply upward. Students may not be familiar with the terms *magnitude* and *direction*. It's not important that they learn the terms at this time, but it is important that they recognize both changes: The ball is going more slowly, and its direction has changed so that its path is less steep than it was at first. Students can see both effects by comparing vectors v_0 and v_1. Vector v_1 is shorter in length and is pointed less steeply upward.

Q2 With each interval of time, the velocity vector points more downward. As the velocity vector moves downward, the ball is not moving upward as steeply as it was at first (and it soon starts to move downward).

Q3 The path of the ball levels off, becoming more horizontal and less vertical. Soon the ball begins to move downward. (In the default state of the sketch, the ball reaches its apex near p_3 and then begins to move downward.)

Q4 If the velocity pattern continues, the ball's direction of travel will be more and more downward at each time interval. This agrees with students' knowledge of how balls fly through the air, but it is also important that they observe how the changes in the velocity vector correspond to the actual movement of the ball.

Q5 The shape of the ball's path looks like an arc of a parabola. The more downward the velocity vector points, and the longer it gets in this direction, the more vertical the ball's path becomes.

Q6 In the original state of the sketch, the ball begins to move downward during the fourth interval. The v_3 vector is the first one to point downward, with the result that after p_3, the ball begins to move downward.

Q7 Answers will vary. If there is no gravity, the velocity vector will not change over time. With no changes in velocity, the ball will continue to travel in the initial direction, and its path will be a straight line. Encourage students to explain their drawings, and make sure they understand that the ball moves in a straight line forever because gravity does not alter the ball's velocity.

Q8 Students have a mental image of how a thrown ball moves through the air, but it's still useful to demonstrate by gently tossing a soft object. The real ball moves through the air in a smooth arc, while the construction is a rough approximation because its position is only measured at intervals, where the segments connect.

Some students may raise a more sophisticated question about the terminal velocity of a ball and point out that the Sketchpad model doesn't allow for air resistance.

Q9 You should shorten both the velocity and gravity vectors by half. You should shorten the velocity vector because the ball should travel less far in a shorter interval of time. Similarly, you should shorten the gravity vector because during a half-second interval the ball's velocity should not change by as much as it does during a full-second interval.

Q10 As it gets smoother, the iterated path becomes more and more parabolic in shape.

EXPLORE MORE

10. Students can also make the parabola and the iterated path match by changing the path. This approach requires modifying the initial position of the ball, the initial velocity of the ball, and the length of the gravity vector. (The gravity vector must remain vertical or the objects won't match.)

5

Algebraic Transformations

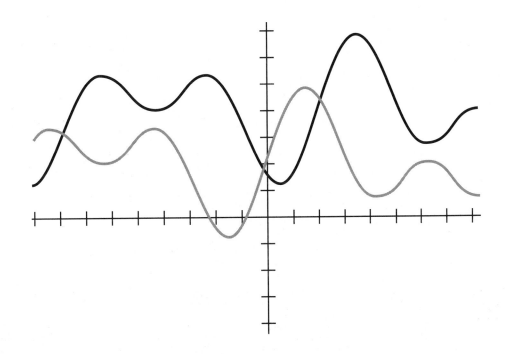

Translating Coordinates

Analytic geometry is a branch of mathematics that uses a coordinate system to study geometry. In this activity you'll do some analytic geometry as you study what happens to the coordinates of points—specifically, the vertices of a triangle—when they're translated.

SKETCH

Choose **Graph | Define Coordinate System.** Then choose **Graph | Snap Points.**

You'll start by creating a coordinate system and drawing the triangle to be translated.

1. In a new sketch, create a new coordinate system with point snapping turned on.

2. All the coordinate measurements in this activity will be integers, so change Sketchpad's preferences to display the coordinates to the nearest integer. Choose **Edit | Preferences,** and on the Units panel set the Precision for Scalars (Slope, Ratio, ...) to **units.**

3. Using the **Segment** tool, draw a triangle whose vertices are grid points.

Select the three points with the **Arrow** tool, and choose **Measure | Coordinates.**

4. Measure the coordinates of the triangle's three vertices. The vertices are labeled *A, B,* and *C.*

5. Construct the triangle's interior by selecting its three vertices and choosing **Construct | Triangle Interior.**

Now you need to tell Sketchpad how to translate the triangle. To do this, you'll construct a segment and use it to define a *translation vector.*

6. Draw a segment from the origin to a point *D* in the plane.

7. Measure the coordinates of point *D.* Drag this new coordinate measurement so that it's away from the others.

Select in order the origin and point *D.* Choose **Transform | Mark Vector.**

8. Mark the vector from the origin to point *D.* A brief animation shows the vector that you've marked.

Now translate the triangle and measure the new coordinates.

9. Translate the triangle by the marked vector by selecting the entire triangle (vertices, sides, and interior) and choosing **Transform | Translate.**

A: (−6, 2)
B: (−4, 3)
C: (−1, 1)

A': (−4, −2)
B': (−2, −1)
C': (1, −3)

D: (2, −4)

10. Measure the coordinates of the new triangle's vertices.

11. Experiment by dragging point D or any of the triangle vertices. Look for a relationship between a point's coordinates, the coordinates of its image under a translation, and the coordinates of point D.

INVESTIGATE

Q1 Where can you drag point D so that the original points and their corresponding image points always have the same y-coordinates? The same x-coordinates?

Q2 When the vector you defined translates the triangle to the left and up, in what quadrant is point D? How do you know?

Q3 Suppose point D has coordinates (s, t). What are the coordinates of the image of a point (x, y) under a translation by (s, t)?

Q4 Quadrilateral $JKLM$ has vertices $J: (-2, -1)$, $K: (-3, 3)$, $L: (-1, 1)$, and $M: (1, 2)$. In its translated image $J'K'L'M'$, J' is at $(1, -2)$, K' is at $(0, 2)$, and L' is at $(2, 0)$. Where is M'?

Q5 In the previous problem, assume that $JKLM$ was translated to $J'K'L'M'$ using the vector in the sketch. What are the coordinates of point D?

EXPLORE MORE

Q6 Using Sketchpad's Measure menu, you can measure angles, point coordinates, segment slope and length, and perimeter and area of the interior. Which of these measurements are preserved (don't change) under a translation? Which are not preserved?

Q7 Triangle UVW has vertices $U: (5, 8)$, $V: (2, -5)$, and $W: (-1, 3)$. Triangle XYZ has vertices $X: (3, 6)$, $Y: (0, -7)$, and $Z: (-2, 0)$. Explain why XYZ cannot be the image of UVW under a translation.

Translating Coordinates

Objective: Students translate points in the coordinate plane and learn the associations between the translation and the image coordinates.

Student Audience: Algebra 1/Geometry/Algebra 2

Prerequisites: Students should understand the (x, y) notation of a point in the coordinate plane. Some previous familiarity with translations is also helpful. The term *translation vector* is used and defined. The term *image* (as in "the coordinates of its image under a translation") is used but not defined, so you may want to discuss this terminology from transformational geometry.

Sketchpad Level: Intermediate. Students start with a blank sketch and construct everything from scratch.

Activity Time: 35–45 minutes. If you have less time or are working with inexperienced Sketchpad users, you can use the sketch **Translate.gsp.** Have students open this sketch, go to the page labeled "After Step 10," and start the activity at step 11.

Setting: Paired/Individual Activity (no sketch required) or Whole-Class Presentation (use **Coordinate Translation Present.gsp**)

You may wish to begin this activity by discussing vectors. What are vectors? How are they different from lines, rays, or segments?

If students finish early, you might suggest that they try translating several times with the same vector or translating using two or more different vectors. How are the final images related to the translation vectors in each case? Students may also try using vectors whose tails aren't attached to the origin.

You might also ask students to extend the activity by exploring the various ways of specifying a translation in Sketchpad. This activity uses a marked vector defined by two points (its tail and head). One alternative is to specify a distance and an angle (a *polar vector*); another is to specify a horizontal distance and a vertical distance (a *rectangular vector*). Any of these three methods of specifying a

vector can be used in Sketchpad, though the focus of this discussion should not be on Sketchpad functionality, but rather on various ways to specify a vector. You could ask students to discuss when they might want to use one method and when they might want to use another.

INVESTIGATE

Q1 If you drag point D along the x-axis, the y-coordinates of a vertex and its image point will be equal. If you drag point D along the y-axis, the x-coordinates of a vertex and its image point will be equal.

Q2 If the vector translates the triangle to the left and up, point D is in the second quadrant. Its x-coordinate is negative (causing the leftward movement) and its y-coordinate is positive (causing the upward movement).

Q3 The image of (x, y) under a translation by (s, t) is the point $(x + s, y + t)$. Note that this answer holds regardless of the signs of x, y, s, or t.

If this question is difficult for students, have them use actual values for x, y, s, and t, all positive at first, and have them look for patterns.

Q4 Point M' will be at $(4, 1)$. The reason is that the translation vector here is $(3, -1)$, three to the right and one down. $(1 + 3, 2 + (-1))$ is $(4, 1)$.

Q5 Point D is at $(3, -1)$.

EXPLORE MORE

Q6 All of the measurements are preserved except for the point coordinates.

Q7 Triangle XYZ cannot be the image of triangle UVW under translation because the three translation vectors aren't the same. The vectors used to get from U to X and from V to Y are both $(-2, -2)$, whereas the vector used to get from W to Z is $(-1, -3)$. Other arguments are possible (for example, the two triangles aren't similar).

This presentation uses coordinates to represent three major components: a triangle in the coordinate plane, a translation vector, and the translated image of the triangle. In Q4–Q6, you will remove one of these things from view and challenge the class to derive it from the two remaining. Do plenty of repetitions, and be sure to give students a look at some negative numbers.

1. Open **Coordinate Translation Present.gsp.** Press *Translate.* A quick animation will show the translation of the triangle by the vector from the origin to point D.

Spend a few minutes discussing the meaning of translation. As you speak, drag points A, B, C, and D. The image will move accordingly. Students will tend to pick up on the geometric relationships better when they see the objects in motion. This would be a good opportunity to acquaint them with the terms *pre-image* ($\triangle ABC$), *image* ($\triangle A'B'C'$), and *translation vector* ($\overrightarrow{OD}$).

Q1 What is the image of point A? (A')

Q2 Concentrate on the coordinates of points A, A', and D. What relationship is there between these numbers? (Give plenty of time for a response. $x_A + x_D = x_{A'}$ and $y_A + y_D = y_{A'}$.)

Q3 What about points B and C and their images? What's the connection? (They have the same association. The differences between corresponding coordinates of a point and its image are defined by the coordinates of point D.)

2. Press *Secret Translation.* This hides all of the objects and moves the pre-image and point D to new locations. By construction, the image must move also.

3. Show the pre-image and point D.

Q4 What are the coordinates of the image points? (Give students time to work it out, and then press *Show Image.*)

4. Press *Secret Translation* again. This time show the pre-image and the image.

Q5 What are the new coordinates of point D? (Press *Show Point D.*)

5. Press *Secret Translations* again. Show D and the image.

Q6 What are the coordinates of the pre-image points? (Press *Show Pre-image.*)

When you think students are ready, turn off the presentation and do the same thing on a board, this time using coordinates only—no sketches.

Rotating Coordinates

In this activity you'll investigate what happens to the coordinates of points after you rotate them about the origin, mainly by multiples of 90°. If you do the Explore More exercises, you'll do the same thing in polar coordinates.

SKETCH AND INVESTIGATE

Choose **Graph | Define Coordinate System** and then **Graph | Snap Points.**

1. In a new sketch, create a new coordinate system with point snapping turned on.

2. Using the **Segment** tool, draw a quadrilateral whose vertices are grid points. Use the **Text** tool to label the vertices *A*, *B*, *C*, and *D*.

3. Select the four vertices consecutively clockwise or counter-clockwise, and choose **Construct | Quadrilateral Interior.**

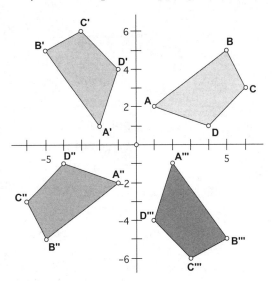

4. Mark the origin of your coordinate system as a center of rotation by selecting it and choosing **Transform | Mark Center.**

To rotate one or more objects, select them and choose **Transform | Rotate.** Enter the angle to use in the dialog box that appears. Click Rotate when the settings are correct.

5. Rotate the entire quadrilateral (points, sides, and interior) by 90°.

6. Drag a vertex of the original quadrilateral, and observe how the image responds.

7. Now rotate the image quadrilateral by 90°. Rotate the next image too. You should have four quadrilaterals on the screen.

Q1 If you were to rotate the fourth quadrilateral, where would its image be?

8. Select point *A* and its three rotated images (its *corresponding* points on the other three quadrilaterals). Choose **Measure | Coordinates.**

Q2 What is the relationship between the coordinates of a point and the coordinates of its image point after a 90° rotation about the origin? Suppose that a point with coordinates (a, b) is rotated by 90° about the origin. What are the coordinates of its image point?

Q3 What is the relationship between the coordinates of a point and the coordinates of its image point after a 180° rotation about the origin? If a point with coordinates (a, b) is rotated by 180° about the origin, what will the coordinates of its image point be?

Q4 If you rotate an object by −90°, it will be rotated clockwise rather than counter-clockwise. Suppose you rotate a point with coordinates (*a*, *b*) by −90°. What will the coordinates of the rotated point be?

Q5 What are the coordinates of *B*? Without doing any other measurements or even looking at the screen, what are the coordinates of the three rotated images of *B*?

EXPLORE MORE

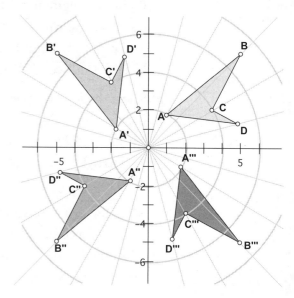

Q6 Using Sketchpad's Measure menu, you can measure angles, point coordinates, a segment's slope and length, and a polygon's perimeter and area. Which of these measurements are preserved (don't change) under a rotation? Which are not preserved?

9. Many people prefer to use a grid for certain kinds of graphing. To get a polar grid, choose **Graph | Grid Form | Polar.** Now once again measure the coordinates of a point and its image points. Then drag the points around to determine the relationship between the coordinates of rotated points.

Q7 Again suppose that a point has coordinates (*a*, *b*), but in this case they are polar coordinates. What are the coordinates of the images when you rotate this point 90° about the origin three times in succession?

Q8 Rotate a point by 45° about the origin. Measure the coordinates of the original point and its image point in both a square coordinate system and a polar coordinate system. In which type of coordinate system is the relationship between the coordinates clearer?

Exploring Algebra 2 with The Geometer's Sketchpad
© 2007 Key Curriculum Press

Rotating Coordinates

Activity Notes

Objective: Students learn some basics of coordinate rotation by rotating figures about the origin by multiples of 90°.

Student Audience: Algebra 1/Geometry/Algebra 2

Prerequisites: Students should understand the (x, y) notation of a point in the coordinate plane. Some previous familiarity with rotation is also helpful. The term *image* (as in "the coordinates of its image point after a 90° rotation") is used but not defined, so you may want to discuss this terminology from transformational geometry.

Sketchpad Level: Intermediate. Students start with a blank sketch and construct everything from scratch. All steps are clearly explained (and extra tips are given below), but if students are unfamiliar with Sketchpad, the activity might take longer than the estimate given.

Activity Time: 30–40 minutes

Setting: Paired/Individual Activity (no sketch required) or Whole-Class Presentation (use **Coordinate Rotation Present.gsp**)

While this activity is progressing, you may want to ask students why only multiples of 90° are being considered. The Explore More section addresses this issue, but students may not get that far and the issue is worth thinking about before then anyway.

A good extension to this activity involves asking students why the relationships in this activity hold. Consider suggesting that students form the triangles defined by the origin, a point, and the point's image.

As another extension, you could discuss whether these relationships will hold if the center of rotation is a point other than the origin.

CONSTRUCTION TIPS

The instructions call for the Snap Points setting. It's not actually necessary to change this, but the coordinate relationships will be easier for students to see if they work with integers.

SKETCH AND INVESTIGATE

Q1 A fourth rotation would map the image to the original pre-image quadrilateral.

Q2 If you switch the x- and y-coordinates of the original point, then change the sign of the new x-coordinate, you get the coordinates of the image point after a 90° rotation about the origin. The coordinates of the image point are $(-b, a)$.

Q3 Both the x- and y-coordinates of the image point are the opposite of those of the original point. The coordinates of the image of point (a, b) are $(-a, -b)$.

Q4 The coordinates of the image point are $(b, -a)$.

Q5 The answer will depend on the coordinates of point B. The coordinates of the four points will have the form $B(a, b)$, $B'(-b, a)$, $B''(-a, -b)$, and $B'''(b, -a)$,

EXPLORE MORE

Q6 Angles, length, perimeter, and area are preserved. Coordinates and slope are not.

Q7 In polar coordinates, the first coordinate represents the distance of a point from the origin. That distance does not change when the origin is the center of rotation. The second coordinate, the angle from the positive x-axis, changes in increments of 90°. Starting with point $A(a, b)$, the images should be in the form $A'(a, a + 90°)$, $A''(a, a + 180°)$, and $A'''(a, a + 270°)$.

Q8 In a square coordinate system, it's difficult to see the relationship between the coordinates of a point and those of its image under a 45° rotation. In a polar coordinate system, the relationship is simple: The r-coordinates are the same, and the θ-coordinate of the image point is that of the original point plus 45°.

To begin this presentation, explain that you intend to investigate some relatively simple rotations. The origin will be the center of rotation, and all rotations will be multiples of 90°.

1. Open **Coordinate Rotation Present.gsp.**

2. Press the Rotate button. This generates rotations of the quadrilateral by angles of 90°, 180°, and 270°. Drag some of the vertices on the red quadrilateral to show how the rotations respond.

3. Take a moment to introduce or review the transformation notation used here. When a point is transformed, a prime (′) symbol is added to the label of its image, which is why all of the point labels on the last image have three prime symbols.

4. Press *Show A*.

Q1 These are the coordinates of point A. What should the coordinates be for A'? The actual answer will depend on the coordinates of A. Given A: (a, b), the coordinates of A' will be $(-b, a)$.

Go through each of the Show buttons one at a time to see all three images of A. Pause each time to ask for predictions. The general form will be as follows:

$$A(a, b), \quad A'(-b, a), \quad A''(-a, -b), \quad A'''(b, -a)$$

5. Press *Show B*.

Q2 What are the coordinates for the three images of point B? Give students time to agree on an answer for all three before revealing the coordinates. They should anticipate the same pattern.

Page 2 is actually the same sketch, but with the settings changed. Walk through the same presentation again, and ask the class to predict coordinates on the rotated points. This progression follows a different, simpler pattern:

$$A(a, b), \quad A'(a, b + 90°), \quad A''(a, b + 180°), \quad A'''(a, b + 270°)$$

All of the θ-coordinates will be reduced to equivalent angles between $-180°$ and $180°$.

If time allows, press *Reset*, hide the coordinates, and drag vertices to start again with a different quadrilateral. Do this on both pages. Try putting the pre-image somewhere other than the first quadrant.

Reflecting in Geometry and Algebra

If you're like most people, you've spent at least a little time looking at yourself in the mirror, so you're already familiar with reflection. In this activity you'll add to your knowledge on the subject as you explore reflection from both geometric and algebraic perspectives.

SKETCH AND INVESTIGATE

1. In a new sketch, use the **Point** tool to draw a point.

2. With the point still selected, choose a color from the **Display | Color** submenu. Then choose **Display | Trace Point.** Use the **Arrow** tool to drag the point around. The trail the point leaves is called its *trace*.

3. If the trace remains on the screen without fading, choose **Edit | Preferences.** On the Color panel, check Fade Traces Over Time and click OK.

To choose the **Line** tool, press and hold the mouse button on the **Straightedge** tool, then drag and release over the **Line** tool in the palette that appears.

4. Using the **Line** tool, draw a line. With the line selected, choose **Transform | Mark Mirror.** An animation indicates that the mirror line has been marked.

5. Using the **Arrow** tool, select the point. Choose **Reflect** from the Transform menu. The point's reflected image appears.

6. Give the new point a different color, and turn on tracing for it as well.

Starting in this step, we'll refer to the two points defining the line as *line points* and the other two points as *reflecting points.*

7. What will happen when you drag one of the reflecting points? Ponder this a moment. Then drag and see. What do you think will happen when you drag one of the line points? Find the answer to this question too.

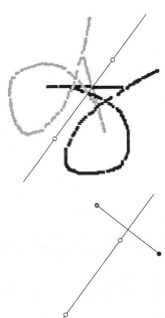

Q1 Briefly describe the two types of patterns you observed in step 7 (one when dragging a reflecting point, the other when dragging a line point).

8. Select the reflecting points; then choose **Display | Trace Points** to toggle off tracing.

9. With the two points still selected, choose **Construct | Segment.** A segment is constructed between the points. Drag the various objects around and observe the relationship between the line and the segment.

Q2 What angle do the line and the segment appear to make with each other? How does the line appear to divide the segment?

FROM GEOMETRY TO ALGEBRA

Now that you've reviewed some geometric properties of reflection, you can apply this knowledge to reflection in the *xy* plane. Start by exploring reflection across the *y*-axis.

10. Click in blank space to deselect all objects. Drag one of the line points so it's near the center of the sketch. With this point selected, choose **Define Origin** from the Graph menu. A coordinate system appears. The selected point is the origin, $(0, 0)$.

11. Deselect all objects; then select the *y*-axis and the other line point (the one that didn't become the origin). Choose **Edit | Merge Point To Axis.** The point now attaches itself to the *y*-axis, which acts as the mirror line.

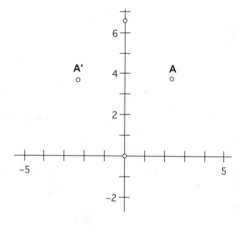

12. Drag one of the reflecting points and consider the coordinates of both points as they move. How do the coordinates of the two points compare?

13. Select both of the reflecting points. Choose **Measure | Coordinates.**

Q3 A point with coordinates (a, b) is reflected across the *y*-axis. What are the coordinates of its reflected image?

14. Now consider the distance between the two reflecting points and how it relates to their coordinates. Make a prediction. Then select the two points and choose **Measure | Coordinate Distance.**

A special challenge is to make sure your answers to this question and Q6 work regardless of what quadrants the points are in.

Q4 A point with coordinates (a, b) is reflected across the *y*-axis. How far is it from its reflected image?

15. Deselect all objects. Then select the point on the *y*-axis that you merged in step 11. Choose **Split Point From Axis.**

The point is split from the *y*-axis.

16. With the point still selected, select the *x*-axis as well. Then choose **Edit | Merge Point To Axis** from the Edit menu. The *x*-axis now acts as the mirror line. Drag one of the reflecting points and observe the various measurements.

Exploring Algebra 2 with The Geometer's Sketchpad
© 2007 Key Curriculum Press

Q5 A point with coordinates (a, b) is reflected across the x-axis. What are the coordinates of its reflected image?

Q6 A point with coordinates (a, b) is reflected across the x-axis. How far is it from its reflected image?

EXPLORE MORE

17. Plot the line $y = x$. Split the point from the x-axis and merge it to the new line.

Q7 A point with coordinates (a, b) is reflected across the line $y = x$. What are the coordinates of its reflected image?

18. Consider the following transformations (each is separate):

    a. Reflect a point over the x-axis, and then reflect the image over the y-axis.

    b. Reflect a point over the y-axis, and then reflect the image over the x-axis.

    c. Rotate a point by 180° about the origin.

Q8 How do these three transformations compare? What would the coordinates of a point (a, b) be after each of these transformations?

Reflecting in Geometry and Algebra

Objective: Students reflect points across the coordinate axes and learn the algebraic associations between the coordinates of the point and its image.

Student Audience: Algebra 1/Geometry/Algebra 2

Prerequisites: Familiarity with the Cartesian plane

Sketchpad Level: Easy/Intermediate. The constructions steps that students do are explained in detail.

Activity Time: 30–40 minutes

Setting: Paired/Individual Activity (no sketch required) or Whole-Class Presentation (use **Reflection Present.gsp**)

Related Activities: Reflecting Functions

This activity works well as a brush-up for students having problems with the coordinate plane, as an introduction to using Sketchpad for both geometry and algebra, and as preparation for function transformation in later activities.

Before starting this activity, it would be a good idea to brainstorm what students already know about reflection. A useful question to discuss—anticipating Q4 and Q6—is "If you stand 3 feet from a mirror, how far do you *appear* to be from your reflected image?"

During the activity you may wish to encourage students to continually connect the geometric and algebraic realms. For example, how do the relationships found in Q4 and Q6 relate to those from Q2?

If your class has done this and the previous two activities, you may want to end this activity by comparing all three transformations—translation, rotation, and reflection. Which transformation was the easiest to understand from an algebraic perspective? Which was the most surprising?

CONSTRUCTION TIPS

3. You may choose instead to leave Fade Traces Over Time unchecked so that traces remain on screen to be examined. In this case, students would periodically need to choose **Erase Traces** from the Display menu to clear traces from their screens.

14. The reason for using **Coordinate Distance** instead of **Length** or **Distance** from the Measure menu is that the scale of the axes can be changed. A unit on the coordinate grid does not necessarily have the same length as the distance measurement unit.

SKETCH AND INVESTIGATE

Q1 Dragging a reflecting point results in a mirror pattern with the two traces mirroring each other across the line. Dragging a line point causes the reflected image point to draw a circle around the other line point. The radius of this circle is the distance between the other line point and the reflecting pre-image.

Q2 The line is the segment's perpendicular bisector, meaning that the angle they form is 90° and the line cuts the segment in half.

FROM GEOMETRY TO ALGEBRA

Q3 $(-a, b)$

Q4 The distance is $|2a|$ or $2|a|$, but give some credit for the answer $2a$. The absolute value signs ensure that the answer will be positive even if a is negative. This is desirable because the distance between two objects is always considered to be non-negative.

Q5 $(a, -b)$

Q6 $|2b|$ or $2|b|$, and maybe $2b$ (See the answer to Q4 above.)

EXPLORE MORE

Q7 The coordinates switch places. The image of a point (a, b) reflected across the line $y = x$ is the point (b, a).

Q8 These three transformations are equivalent. The coordinates of (a, b) after any of the transformations are $(-a, -b)$.

Reflecting in Geometry and Algebra

1. Open **Reflection Present.gsp.** Drag point *P* as you explain that *P'* is its reflection across the blue line. Also drag the points that define the line. Experiment with the *Show Traces* button to emphasize the reflection properties.

2. Construct line segment *PP'*.

Q1 What is the relationship between this line segment and the mirror line? (The mirror line is the perpendicular bisector of the line segment.)

3. Open page 2. This is essentially the same construction moved to the coordinate plane. Now you can control the mirror line only by using the action buttons. Press *Reflect Over x-axis* to align the mirror line with the *x*-axis.

All of the answers on this page are with respect to coordinates (*a, b*) for point *P*.

Q2 The coordinates of point *P* are on the screen. Ask, "Knowing what you do about the properties of reflection, what are the coordinates of *P'*?" (For $P(a, b)$, the coordinates of *P'* are $(a, -b)$. Press *Show Coordinates of P'*.)

Q3 What is the distance *PP'*? ($|2b|$)

4. Press *Hide Coordinates of P'*, *Hide PP'*, and *Reflect Over y-axis*.

Q4 Now what are the coordinates of *P'*? $(-a, b)$

Q5 What is the distance *PP'*? $|2a|$

5. Hide the measurements again, and press *Reflect Over y = x*. This one is more difficult, so take it slowly.

Q6 What are the coordinates of *P'*? (b, a)

Q7 What is the distance *PP'*? $|\sqrt{2}(a - b)|$

6. Press *Reflect Over x-axis*.

Q8 Here it is reflected on the *x*-axis again. Ask, "What will you get if you reflect this image, *P'*, across the *y*-axis?" (The second image is $P''(-a, -b)$, and $PP'' = 2\sqrt{a^2 + b^2}$.)

7. After plenty of discussion, press *Show Second Reflection*, *Show Coordinates of P''*, and *Show PP''*.

Q9 Ask, "What happens if you reflect it on the *y*-axis first and then the *x*-axis?"

8. Leaving all of the objects still showing, press *Reflect Over y-axis*. Both of the mirror lines will rotate together, and as this happens, students will see that *P''* does not move at all. This should lead to a general conclusion about reflections on two perpendicular lines.

Stretching and Shrinking Coordinates

In geometry dilation is defined in terms of moving toward or away from a center point. On the coordinate plane we can define an additional transformation in which an object is moved toward or away from a line, usually one of the coordinate axes. Such transformations are usually called *shrinks* and *stretches*. In this activity you'll create shrinks and stretches and investigate their properties.

STRETCH AND SHRINK

First you'll use a stretch to move a point away from the *x*-axis.

1. Open **Stretch Shrink Coords.gsp.** The sketch contains a coordinate system, a polygon, and a slider labeled *a*.

Use the **Text** tool to label the point.

2. Construct a point using the **Point** tool. Label it *A* and measure its *x*- and *y*-coordinates.

To measure the *x*-coordinate, select the point and choose **Measure | Abscissa (x).** Use a similar method for *y*.

Q1 Which of these two measurements tells you how far the point is from the *x*-axis?

Q2 How could you change the point's coordinates to define a new point that's twice as far away from the *x*-axis?

Enter the value y_A into the calculation by clicking on the y_A measurement in the sketch.

3. Use Sketchpad's Calculator to multiply the measured *y*-coordinate by 2. To do this, choose **Measure | Calculate** and calculate $2y_A$.

4. Construct the new point by selecting in order the values x_A and $2y_A$ and then choosing **Graph | Plot As (x, y).** This new point is called the *image*, and the original point *A* is called the *pre-image*.

5. Drag pre-image *A* and observe the behavior of the image.

Q3 How does the image behave when the pre-image is near the *x*-axis? How does it behave when the pre-image is below the *x*-axis?

To edit the calculation, double-click it. Then change the expression to put *a* (the value of the slider) in place of the 2.

6. Edit the calculation $2y_A$ to make it $a \cdot y_A$, so that it multiplies by the value of the slider instead of by the constant 2.

7. Drag the slider to set its value to 0.50.

$x_A = 0.70$
$y_A = 1.80$
$a \cdot y_A = 0.90$

$a = 0.50$

Q4 Drag the pre-image point. How does the image point behave now? Is this a stretch or a shrink?

Now you'll attach the point to the polygon border and use it to apply the same transformation to every point of the polygon.

8. Merge point *A* to the hexagon by selecting both the point and the hexagon interior and choosing **Edit | Merge Point To Hexagon.**

Q5 Drag point *A* around the hexagon. How does the image point behave now?

9. To construct the locus of the image as pre-image *A* moves around the hexagon, select both points and choose **Construct | Locus.**

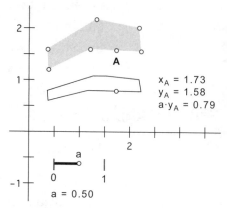

$x_A = 1.73$
$y_A = 1.58$
$a \cdot y_A = 0.79$

Q6 Drag the slider so *a* = 2. How does this change the locus? What does the locus look like if you make *a* = 1?

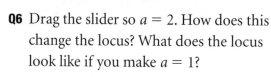
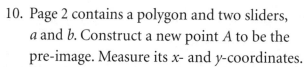

$a = 0.50$

10. Page 2 contains a polygon and two sliders, *a* and *b*. Construct a new point *A* to be the pre-image. Measure its *x*- and *y*-coordinates.

11. Use Sketchpad's Calculator to compute the new coordinates: $b \cdot x_A$ and $a \cdot y_A$.

12. Construct the image point by selecting the two calculations in order and choosing **Graph | Plot As (x, y).**

13. As you did before, attach point *A* to the polygon, and then construct the locus of the image point as pre-image *A* moves around the polygon.

14. Experiment by dragging the sliders and observing the effect on the locus.

Q7 Set the sliders so *a* = 2 and *b* = 2. Describe the shape of the locus compared to the original polygon.

Q8 Describe the shape of the locus, relative to the original polygon, for each combination of slider values below:

    a. *a* = 0.50 and *b* = 0.50   b. *a* = 1 and *b* = 1

    c. What conclusion can you draw when the two sliders have equal values?

Q9 Now describe the results for these slider values:

    a. *a* = 1 and *b* = 3   b. *a* = 0.50 and *b* = 1

    c. *a* = 2 and *b* = 3

EXPLORE MORE

Q10 Describe the transformations that result from these slider values:

    a. *a* = 1 and *b* = −1   b. *a* = −1 and *b* = 1

    c. *a* = −0.5 and *b* = −0.5

What can you conclude about stretches and shrinks by negative values?

© 2007 Key Curriculum Press

Stretching and Shrinking Coordinates

Objective: Students multiply the coordinates of a point to stretch or shrink a polygon toward or away from the axes. They multiply by different numbers and observe the resulting transformations. This activity lays the groundwork for stretching and shrinking functions.

Student Audience: Algebra 2

Prerequisites: It is helpful if students are familiar with geometric transformations, and particularly with dilation.

Sketchpad Level: Intermediate.

Activity Time: 25–35 minutes

Setting: Paired/Individual Activity (use **Stretch Shrink Coords.gsp**) or Whole-Class Presentation (use **Stretch Shrink Coords Present.gsp**)

Related Activities: Stretching and Shrinking Functions

STRETCH AND SHRINK

The informal definition of stretches and shrinks in the first paragraph of this activity describes them as moving points toward or away from the axes, suggesting that only a single object is involved. The activity itself uses the terms pre-image and image, implying two separate objects. This apparent contradiction—whether a transformation involves changing a single object or creating a second object related to the first—is confusing and merits class discussion. In fact, we often use both ways of talking about transformations, because a transformation can be viewed in a static way (with a pre-image giving rise to a separate image) or in a dynamic way (with a single object changing from one form to another). Each way of describing transformations is useful; encourage students to become comfortable with both.

Q1 The y-value tells how far the point is from the x-axis. (The x-value tells you how far it is from the y-axis.)

Q2 A point twice as far from the x-axis must have a y-value twice as great, so you should multiply the y-value by 2.

Q3 The image point always stays twice as far from the x-axis, no matter where you move the pre-image.

Q4 The image is now only half as far from the x-axis as the pre-image. This is a shrink.

Q5 The image point traces out a path related to the border of the polygon, but with a vertical shrink. This path has the same width as the polygon, but is only half as high.

Q6 When $a = 2$, the image point traces out a path related to the border of the polygon, but with a vertical stretch. This path is as wide as the polygon, but is twice as high. When $a = 1$, the image coincides with the pre-image.

Q7 When both a and b are 2, the image is twice as large as the original polygon, both vertically and horizontally. This is equivalent to a geometric dilation by a factor of 2.

Q8 a. When $a = 0.50$ and $b = 0.50$, the image is half the size of the original, both vertically and horizontally.

b. When $a = 1$ and $b = 1$, the image coincides with the original polygon.

c. When the two sliders have equal values, the image is transformed the same way vertically and horizontally, so the image has the same shape as the original, although it may be larger or smaller.

Q9 a. When $a = 1$ and $b = 3$, the image is stretched by a factor of 3 horizontally, but not at all vertically, so it's three times as wide as the original and the same height as the original.

b. When $a = 0.50$ and $b = 1$, the image is half as high as the original and the same width.

c. When $a = 2$ and $b = 3$, the image is twice as high as the original and three times as wide.

EXPLORE MORE

Q10 a. When $a = 1$ and $b = -1$, the image is the same size as the original, but is reflected across the y-axis.

b. When $a = -1$ and $b = 1$, the image is the same size as the original, but is reflected across the x-axis.

c. When $a = -0.5$ and $b = -0.5$, the image is half the size as the original, both vertically and horizontally. However, it is rotated by 180° about the origin. (This rotation is equivalent to two reflections, across each of the axes.)

Stretches and shrinks by negative values result in reflection about the corresponding axis.

Stretching and Shrinking Coordinates

Presenter Notes

This presentation lays the groundwork for using stretches and shrinks to transform functions. First remind students how dilation works in geometry. Explain that the transformations involved in this presentation are different, because dilation involves equal stretches or shrinks both vertically and horizontally, but here they will see stretches and shrinks that are different in the two directions.

STRETCH AND SHRINK

First you'll shrink or stretch to move a point toward or away from the *x*-axis.

1. Open **Stretch Shrink Coords Present.gsp.** The sketch contains point *A*.

2. Press the buttons to measure *A*'s coordinates and to multiply the *y*-value by 2.

3. Press *Construct Image of A* to plot a new point using the calculated *y*-value.

Q1 Drag point *A* to a variety of different locations, and ask students to describe the behavior they observe.

Now make the multiplier adjustable by using the *a* slider.

4. Press *Show Slider* to make the slider visible, and press *Use Slider as Multiplier* to use it in the calculation in place of the constant 2.

Q2 Drag point *A* to several different locations, and ask students to describe how the image behaves. Then change the value of the slider and drag *A* again.

5. On page 2 the pre-image point *A* has been attached to a polygon. Drag the point and ask students to observe the behavior of its image *A'*. Press *Show Polygon Image* to show the path of the image point.

Q3 Adjust the *a* slider. Ask students to describe the shape and behavior of the image for various values of *a*.

6. On page 3 are two sliders, allowing you to transform the *x*- and *y*-values separately. Press the first few buttons to show the various elements and to show the image of the polygon.

Q4 Ask students to predict the image for each combination of slider values listed in the sketch. Then adjust the sliders to determine how accurate the predictions were. (All the combinations on this page have *a* = *b*, so the image is a similar shape to the pre-image.)

Pages 4 and 5 contain similar lists of slider values. If students can predict most of these correctly, they have a good sense of how stretches and shrinks work.

Transforming Coordinates

There are several Sketchpad activities in which you transform a point geometrically and then predict and observe the effect on the point's algebraic coordinates. In this activity you'll do the reverse: You'll transform the coordinates and observe the geometric effect on the plotted images.

GETTING STARTED

The coordinate system appears automatically.

1. In a new sketch, construct a point and measure its coordinates using **Measure | Abscissa (x)** and **Measure | Ordinate (y).**

To put a measurement like x_A into a calculation, click the measurement in the sketch.

To change the label of a calculation, double-click it using the **Text** tool. Put the subscript in square brackets: x[A'] for $x_{A'}$.

2. To transform the coordinates algebraically, choose **Measure | Calculate,** and use the Calculator to define these two values:

$$x_{A'} = x_A + 3 \quad y_{A'} = y_A - 5$$

3. Plot the point determined by these transformed coordinates by selecting them in order and choosing **Graph | Plot As (x, y).** Label the new point A'.

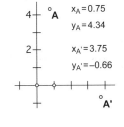

Q1 The two formulas define the coordinates of image point A'. Drag A and describe how the transformation behaves. What type of transformation is this?

You can understand the transformation more easily by applying it to more than one point. Here you will apply it to a polygon.

Holding the Shift key while you click keeps all the points selected.

4. To construct the polygon, hold down the Shift key and use the **Point** tool to construct several points to be the vertices. While the points are still selected, choose **Construct | Polygon Interior.** Drag the vertices to give the polygon whatever shape you want.

5. To attach point A to the polygon boundary, select both A and the polygon, and choose **Edit | Merge Point To Polygon.**

6. Drag point A around the polygon and observe the path of A'. Then construct the locus of A' by selecting points A and A' and choosing **Construct | Locus.**

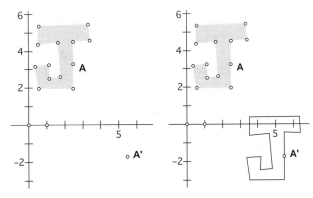

Now you can see the transformation more clearly. Drag the polygon vertices and watch how the image moves. Does this confirm your response to Q1?

Q2 You can change the definitions for the ordered pair $(x_{A'}, y_{A'})$. Double-click on a calculation in order to edit it. For each pair of definitions below, draw on your paper a prediction of what the polygon image will look like. Then change the coordinate definitions to check your prediction.

a. $(-x_A, y_A)$ b. $(-x_A, -y_A)$ c. $(-y_A, x_A)$

d. (y_A, x_A) e. $(2x_A, 2y_A)$ f. $(3x_A, y_A)$

Q3 Here are some transformations that were not included in the list above. State definitions for $(x_{A'}, y_{A'})$ that will create these transformations:

a. Reflect on the line $y = -x$.

b. Rotate 90° clockwise about the origin.

MORE DIFFICULT TRANSFORMATIONS

7. The coordinate definitions below produce a dilation by ratio 3 with respect to the point (2, 4). Edit the coordinates accordingly to confirm that this is correct.

$$x_{A'} = 2 + 3(x_A - 2) \quad y_{A'} = 4 + 3(y_A - 4)$$

Q4 Using a method similar to that in step 7, find the coordinate definitions for the following transformations:

a. Rotate 90° counter-clockwise about the point $(-5, 2)$.

b. Reflect across the line $y = 6$.

EXPLORE MORE

Q5 Construct a point in an arbitrary location, and use it as the center of dilation or rotation. The trick is to measure the coordinates of the center point, then use them in the calculations that define the coordinates of A'.

Q6 You can also use a slider or a parameter to define the ratio of dilation. Give that a try.

Q7 Each of the coordinate definitions described here contained only one variable (either x_A or y_A). Try a definition like $x_{A'} = x_A + 0.5y_A + 4$, $y_{A'} = y_A - 1$. Describe your results.

Exploring Algebra 2 with The Geometer's Sketchpad
© 2007 Key Curriculum Press

Transforming Coordinates

Objective: Students perform elementary transformations in the coordinate plane by applying mapping formulas to point coordinates.

Student Audience: Algebra 1/Algebra 2

Prerequisites: It would be best if students first have experience graphically rendering translations, reflections, right angle rotations, and dilations in the coordinate plane.

Sketchpad Level: Intermediate

Activity Time: 30–40 minutes

Setting: Paired/Individual Activity (no prepared sketch) or Whole-Class Presentation (use **Coordinate Transformation Present.gsp**)

The presentation document has constructions for all of the transformations in Q1 and Q2. If there is a need to save time, or if the students have not acquired the necessary Sketchpad skills, they can use this prepared document.

GETTING STARTED

2. If students are not already familiar with the Calculator, demonstrate the process of clicking on measurements in the sketch to insert them into a calculation.

Q1 This is a translation by vector $(3, -5)$.

Q2 a. Reflection across the y-axis

b. Rotation by 180° about the origin

c. Rotation by 90° counter-clockwise about the origin

d. Reflection across the line $y = x$

e. Dilation with respect to the origin by ratio 2

f. Horizontal stretch from the y-axis by ratio 3

Q3 a. $x_{A'} = -y_A$, $y_{A'} = -x_A$

b. $x_{A'} = y_A$, $y_{A'} = -x_A$

MORE DIFFICULT TRANSFORMATIONS

Q4 a. $x_{A'} = -5 - (y_A - 2)$, $y_{A'} = 2 + (x_A + 5)$

b. $x_{A'} = x_A$, $y_{A'} = 12 - y_A$

EXPLORE MORE

Q5 Examples of dilation and rotation about arbitrary points are shown on pages B and C of the presentation file, **Coordinate Transformation Present.gsp.**

Q6 Pages B and C of the presentation file also contain sliders to define the ratio or angle.

Q7 The example given in the question describes a translation combined with a shear, and is illustrated on page D of the presentation file. (That page also contains a slider to allow you to adjust the shear.) Calculations that use linear combinations of x_A and y_A make possible the full set of affine transformations, including translation, rotation, reflection, dilation, stretching, and shear. An *affine* transformation is one that preserves collinearity and ratios of distances. In other words, any affine transformation of a segment with a midpoint results in another segment with a midpoint: The segment remains straight (collinearity) and the midpoint remains in the middle (ratio of distances).

Other activities are available to show the effects that geometric transformations have on coordinates. In this presentation you will turn it around by transforming the coordinates in order to transform the image.

1. Open **Coordinate Transformation Present.gsp.** Drag point A and ask students to observe its coordinates.

Q1 Point out the two calculations in blue. The first one adds 3 to x_A, and the second one subtracts 5 from y_A. These will be the coordinates of a new point. Where will it appear in relation to point A? (3 units to the right and 5 down)

2. Select the blue coordinates in order and choose **Graph | Plot As (x, y).** Label the new point A'. Drag point A again so students can see that A' moves with it.

Q2 Press the *Show Polygon* button. What would the result be if you added 3 to the x-coordinate and subtracted 5 from the y-coordinate of every point on the polygon? (The entire polygon would be translated.)

3. Select point A and the polygon interior. Choose **Edit | Merge Point To Polygon Interior.** Animate or drag point A so that it moves around the polygon.

Q3 As point A is moving, what is A' doing? (It is tracing a congruent polygon.)

4. Select points A and A' and choose **Construct | Locus.** The translated polygon image appears. Drag vertices of the pre-image polygon to help show the relationship.

The pages labeled (a)–(f) correspond to Q2 from the student activity. In each case point A is already attached to the polygon. The transformation has been constructed, but it is hidden. Stop at each page and challenge the class to describe the coordinate transformation. Use the *Show* buttons to show the results.

Q4 Move to the next page. Where will the polygon image be? Press the *Show Image* button to see the transformed polygon.

Q5 Ask for a precise geometric description of the transformation. Press the *Show Description* button to check the responses.

The pages labeled A, B, C, and D are considerably more complex. The first three involve transformations with respect to points other than the origin, and page D involves a shear. A slider controls the scalar r, and a dial controls angle q.

The final page, labeled "Custom," may be used to create your own transformations. Describe a transformation and ask students how to define it. Edit the blue coordinate calculations according to their directions.

© 2007 Key Curriculum Press

Translating Functions

When you analyze a function, it helps to be able to picture its graph without necessarily drawing it. This is not easy if the function definition is complicated, but you may be able to recognize it as a translation of a simpler graph.

ADD TO THE FUNCTION

1. Open **Translating Functions.gsp.** You will see the function graph $y = f(x)$. The function itself is not important to this investigation. It's just a curve with plenty of ups and downs, giving it a distinctive shape. Point P is attached to the x-axis. The measurement x_p is the x-coordinate of P.

In this section you will transform the function by adding a constant to the value of the function itself.

Enter the function f by clicking the function definition on the screen.

2. Choose **Measure | Calculate** and calculate $f(x_p)$.

3. Select in order x_p and $f(x_p)$. Choose **Graph | Plot As (x, y).** A new point appears. Label it A.

Q1 Drag point P along the x-axis. What is the path of point A?

Now you will see what happens graphically when you add a constant to the function.

4. Using the Sketchpad Calculator again, compute the value $f(x_p) + 4$. Plot point B with coordinates $(x_p, f(x_p) + 4)$.

Q2 Where is point B in relation to point A? Is this relationship the same for all values of x? Test this by dragging point P.

Q3 Describe the path of B as you drag point P.

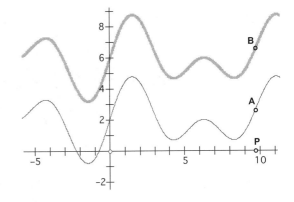

5. Select B and choose **Display | Trace Plotted Point.** Drag P again to trace the path of B.

Q4 The curve you have just traced is $y = f(x) + 4$. Describe it as a translation of $y = f(x)$.

To erase the old traces, press the *Erase Traces* button at the bottom of the screen.

6. Double-click the calculation $f(x_p) + 4$. Change it to $f(x_p) - 7$. Drag point P and observe the path of B again. Try some other constants. Double-click the original function to change its definition. Make up a function of your own.

Q5 From your observations, what general conclusion can you draw about the shape of a graph in the form $y = f(x) + k$?

SUBTRACT FROM THE ARGUMENT

You just saw what happens when you add a constant to the function. In this section you will subtract a constant from the argument (x).

7. Go to page 2. This should look familiar. It's the same as page 1.

8. Calculate $x_p - 2$ and $f(x_p - 2)$. Plot the point $(x_p - 2, f(x_p - 2))$. Label the point A.

Q6 Where does point A appear?

9. Plot the point $(x_p, f(x_p - 2))$. Label it B.

Q7 Where does point B appear with respect to point A? Explain why this is so. Drag point P to confirm that this relationship is always true.

Q8 As you drag point P, changing x, what is the path of point B? Trace point B to confirm this.

10. Experiment with subtracting other constants (positive and negative) from the argument. Try several different functions.

Q9 From your observations, what general conclusion can you make about the shape of a graph in the form $y = f(x - h)$?

SUMMARY

11. Open page 3. The two sliders control the parameters h and k.

12. Choose **Graph | Plot New Function.** Define the function $g(x) = f(x - h) + k$.

13. Drag each slider in turn and observe its effect.

Q10 Describe the graph of $y = f(x - h) + k$ as a transformation of $y = f(x)$.

Translating Functions

Objective: Students translate function graphs vertically and horizontally by adding constants to the function and the function argument.

Student Audience: Algebra 2

Prerequisites: Students must be familiar with function notation and graphing.

Sketchpad Level: Intermediate. Most of the work involves editing calculations and function definitions.

Activity Time: 20–30 minutes

Setting: Paired/Individual Activity (use **Translating Functions.gsp**) or Whole-Class Presentation (use **Translating Functions Present.gsp**)

ADD TO THE FUNCTION

Q1 The path of A is $y = f(x)$, the function plot.

Q2 Point B is four units above A. This relationship holds no matter where point P is.

Q3 Since B is always four units above A, and A is always on the function graph, the path of B has the same shape as the function graph and is four units higher.

Q4 The graph of $y = f(x) + 4$ is the graph of $y = f(x)$ translated four units upward.

Q5 For any function $f(x)$ and any constant k, the graph of $y = f(x) + k$ is the same as that of $y = f(x)$ translated k units upward. If $k < 0$, the translation is downward.

SUBTRACT FROM THE ARGUMENT

Q6 Point A falls on the function graph and is two units to the left of point P.

Q7 Points A and B are at the same height because their y-coordinates are the same. The x-coordinate of B is two units greater than that of A, so B is two units to the right of A.

Q8 Since B is always two units to the right of A, and A is always on the function graph, the path of B has the same shape as the function graph and is two units to the right of the function graph.

Q9 For any function $f(x)$ and any constant h, the graph of $y = f(x - h)$ is the same as the graph of $y = f(x)$ translated h units to the right. If $h < 0$, the translation is leftward.

SUMMARY

Q10 For any function $f(x)$ and constants h and k, the graph of $y = f(x - h) + k$ is the same as $y = f(x)$ translated h units to the right and k units upward.

In this presentation you will demonstrate the function translations of a graph in the form $y = f(x - h) + k$ by showing the vertical and horizontal translation components separately.

1. Open **Translating Functions Present.gsp.** Drag point P to show that it controls the measurement x_p. Drag the slider to show changes to parameter k.

2. Press *Show A.* Point A has coordinates $(x_p, f(x_p))$, so it must fall on the graph.

Q1 Drag the slider so that $k = 3$. For now, you are interested in seeing what happens when you add some constant to the function. If you plot the point $(x_p, f(x_p) + k)$, where will it fall? (It will be 3 units above point A.)

3. Press *Show B.* Drag P to show that B is always three units higher.

Q2 What is the path of point B as you drag P left and right? (Its path is a translation of the graph of $y = f(x)$, 3 units upward.)

4. Select point B and choose **Display | Trace Plotted Point.** Drag point P to trace the path.

Q3 What is the equation of this curve? ($y = f(x) + 3$. Show it with other values of k by dragging slider k to a new value and then dragging P again.)

5. Go to page 2 and show point A. This time it has coordinates $(x_p - h, f(x_p - h))$. This again places it on the graph.

Q4 Drag the slider so that $h = 4$, and plot $(x_p, f(x_p - h))$. Where will it fall? (It will be 4 units to the right of point A.)

6. Press *Show B* and drag point P left and right.

Q5 What is its path? (It is a translation of $y = f(x)$, 4 units to the right.)

7. Trace point B to demonstrate the shape of the path.

Q6 What is the equation of its path? ($y = f(x - h)$)

8. Page 3 has sliders for both h and k. The combined graph $y = f(x - h) + k$ is plotted. There are Show/Hide buttons for it. Try several different combinations for the two parameters, and challenge students to describe the translation before you reveal it in the sketch.

9. Double-click the definition for $f(x)$ to edit it. Try several different definitions. Students need to understand that the nature of the parent function has nothing to do with the translation principle.

Reflecting Functions

If you explored what happens when you reflect points across the *x*- and *y*-axes, you learned that the *x*-coordinates of a point and its reflection across the *y*-axis are the opposite of each other. Similarly, the *y*-coordinates of a point and its reflection across the *x*-axis are opposites. These rules allow you to reflect points across the axes. But what if you want to reflect an entire *function plot* across one of the axes?

COMPARE REFLECTIONS

To begin, plot a function and create two other functions related to the first.

In the New Function Calculator, use the ^ key to enter exponents.

1. In a new sketch, plot $f(x) = x^2 - 5x + 5$ using **Graph | Plot New Function.**

To enter *f* into a new function equation, click its equation in the sketch.

2. The **Graph | New Function** command allows you to enter a function without plotting it. Use this command to create the function $g(x) = -f(x)$. Then create the function $h(x) = f(-x)$.

The graph of one of these new functions will be the reflection of the original graph across the *y*-axis (labeled *A* in the figure below); the graph of the other will be the reflection across the *x*-axis (labeled *B* in the figure below).

Q1 Do you think that $g(x)$ will match plot *A* or plot *B*? Which plot will $h(x)$ match? Explain why you paired them up the way you did.

Q2 Plot the two new functions (first one, then the other, so you know which is which). Was your pairing correct?

To plot a function, select its equation and choose **Graph | Plot Function.**

3. To make them easy to identify, choose different colors for $f(x)$, $g(x)$, and $h(x)$. (Use **Display | Color,** and color both the equation and the plot.) Also make the plot of $f(x)$ thick. (Use **Display | Line Weight | Thick.**)

Q3 If $x = 2$, what is $f(x)$? Write the result as an ordered pair.

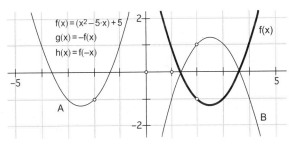

To construct the point, choose the **Point** tool and click on the Function plot.

4. Construct a point on the function plot of *f* at the correct position for this ordered pair. Then construct a corresponding point on the function plot of *g*, and another on the function plot of *h*.

Q4 What are the coordinates of the corresponding point on $g(x)$? Explain why this makes sense algebraically, based on the definition that $g(x) = -f(x)$.

Q5 What are the coordinates of the corresponding point on $h(x)$? Explain why this makes sense algebraically, based on the definition that $h(x) = f(-x)$.

5. Double-click the equation for $f(x)$. In the Edit Function dialog box that appears, change $f(x)$ into some other function—perhaps a different quadratic, an absolute value function, or another polynomial like $f(x) = x^3 - x^2$.

Q6 How does the new $f(x)$ relate to the new graphs of $g(x)$ and $h(x)$?

Q7 Fill in the blanks:

    a. The graphs of $y = f(x)$ and $y = -f(x)$ are reflections of each other across _____ .

    b. The graphs of $y = f(x)$ and $y = f(-x)$ are reflections of each other across _____ .

EXPLORE MORE

Q8 From studying transformations, you've probably learned that reflecting a point over the x-axis, then reflecting that image point over the y-axis gives the same result as reflecting first over the y-axis, then over the x-axis (and that both are equivalent to a 180° rotation). Can you devise a function based on any $f(x)$ that, when plotted, transforms the plot of $f(x)$ twice, in this way?

Q9 An *even function* is one for which $f(-x) = f(x)$. In other words, the plots of $y = f(-x)$ and $y = f(x)$ are the same. An *odd function* is one for which $f(-x) = -f(x)$. In other words, the plots of $y = f(-x)$ and $y = -f(x)$ are the same. Edit $f(x)$ in your sketch to determine which of the following functions are even, which are odd, and which are neither:

    a. $f(x) = x^2 + 3$ b. $f(x) = x^3$

    c. $f(x) = x^3 + 3$ d. $f(x) = x^6 + 3x^4 - 2x^2$

    e. $f(x) = x^5 + 3x^3 - 8x$

Q10 When you reflect a point across the line $y = x$, its x- and y-coordinates are switched. Here's a way of switching the x- and y-values for an entire plot. Choose **Graph | Plot New Function.** Choose **x = f(y)** from the Equation pop-up menu. Enter $f(y)$ (by clicking first on the equation of $f(x)$ in the sketch, then on y in the Calculator keypad), and click OK. Does the resulting graph appear to be a reflection of the original graph across $x = y$? Experiment with several different equations for $f(x)$ to see how a graph relates to its reflection across $y = x$.

Exploring Algebra 2 with The Geometer's Sketchpad
© 2007 Key Curriculum Press

Reflecting Functions

Objective: Students reflect function plots across the x- and y-axes and explore connections between algebraic and geometric transformations.

Student Audience: Algebra 1/Algebra 2

Prerequisites: Students should be comfortable with function notation and function plotting (graphing parabolas, for example). They should also have a general understanding of geometric reflection and transformations.

Sketchpad Level: Intermediate

Activity Time: 15–30 minutes

Setting: Paired/Individual Activity (no sketch required) or Whole-Class Presentation (use **Reflecting Functions Present.gsp**)

Related Activities: Reflecting in Geometry and Algebra, Transforming Odd and Even Functions

A key goal of this activity is for students to understand the notation $y = f(x)$. Specifically, it should help them understand that the part inside the parentheses refers to the input (the x-value), and that using the opposite of x as the input alters its horizontal aspect, so the graph of $f(x)$ is reflected across the y-axis. By contrast, the entire expression $f(x)$ refers to the output (the y-value), and taking the opposite of the output affects the vertical aspect of the graph, so the graph of $f(x)$ is reflected across the x-axis. Students will find this easier to grasp if they start out with a good understanding of function notation and transformations.

COMPARE REFLECTIONS

Q1 Answers will vary. The important thing is that students make a guess and try to explain why.

Q2 Try asking a few students who got the correct pairing to explain their guesses, even if they were based only on a hunch. Also remember to congratulate those students who admit that they guessed wrong; emphasize that we learn best if we're willing to make mistakes.

Q3 If $x = 2$, $f(x) = -1$. The ordered pair is $(2, -1)$.

Q4 The corresponding point on g is $(2, 1)$. This makes sense because the value of $g(x)$ is the opposite of the value of $f(x)$, and 1 is the opposite of -1.

Q5 The corresponding point on h is $(-2, -1)$. By definition, $h(x) = f(-x)$. Therefore, $h(-2) = f(2)$.

Q6 Students should observe that, no matter how they change the function, $g(x)$ still reflects $f(x)$ across the x-axis and $h(x)$ across the y-axis.

Q7 a. The graphs of $y = f(x)$ and $y = -f(x)$ are reflections of each other across the x-axis.

    b. The graphs of $y = f(x)$ and $y = f(-x)$ are reflections of each other across the y-axis.

EXPLORE MORE

Q8 The function $y = -f(-x)$ does the trick. The negative sign outside of $f(x)$ reflects f across the x-axis, and the one on the inside reflects it across the y-axis, as explained in the previous answer.

Q9 a. $f(x) = x^2 + 3$ is even.

    b. $f(x) = x^3$ is odd.

    c. $f(x) = x^3 + 3$ is neither.

    d. $f(x) = x^6 + 3x^4 - 2x^2$ is even.

    e. $f(x) = x^5 + 3x^3 - 8x$ is odd.

Q10 The plot of $x = f(y)$ is not necessarily equivalent to the plot of the inverse function $y = f^{-1}(x)$. To make an inverse function, you may need to restrict the domain of $f(x)$.

In this presentation students will discover how the graphs of $y = f(x)$ and $y = -f(x)$ are reflections of each other across the *x*-axis, and how the graphs of $y = f(x)$ and $y = f(-x)$ are reflections of each other across the *y*-axis. They will also learn how to identify functions as odd or even by observing their reflections.

1. Open **Reflecting Functions Present.gsp.** Press *Show f(x)*, and ask students to identify the type of function and the shape it plots.

Q1 Press *Show g(x) and h(x).* Tell students that *g* and *h* are both reflections of *f(x)*; then ask them to predict which of the two is a reflection across the *x*-axis and which is a reflection across the *y*-axis.

Q2 Ask students for an ordered pair that belongs to *f(x)*. Then ask them for the corresponding ordered pair for *g*, and the corresponding ordered pair for *h*. (For instance, if the ordered pair for *f* is (2, −1), the corresponding pair for *g* must have the opposite *y*-value—(2, 1)—and the corresponding pair for *h* must have the opposite *x*-value—(−2, −1).)

Q3 Ask, "How does the placement of the negative sign affect where the ordered pair will be graphed?" (If the ordered pair has the opposite *x*-value, it will be plotted on the opposite side of the *y*-axis. If it has the opposite *y*-value, it will be plotted on the opposite side of the *x*-axis.)

Q4 Press *Show Plots of g and h,* and ask students to use their predictions to match up the plots with their corresponding functions. (Function *g* matches plot *A*, and function *h* matches plot *B*.)

Q5 Press *Show Pairing* and ask students to evaluate their predictions.

2. Use the second set of functions to reinforce what students have just learned.

Algebraically, for an even function, $f(x) = f(-x)$. For an odd function, $-f(x) = f(-x)$.

3. Go to the "Odd or Even?" page. Explain that for an *even* function, the original function and its reflection across the *y*-axis are identical. For an *odd* function, the reflection across the *x*-axis is identical to the reflection across the *y*-axis.

Q6 For each of the five functions, first press the Show button for that function. Then ask students to decide whether the function is odd, even, or neither.

Q7 Press *Reflect Across x-axis,* and then *Reflect Across y-axis.* Ask students whether their hypotheses were correct. Ask, "How do these reflections help you tell?"

4. Repeat for the remaining functions. For each function, have students decide whether it's odd, even, or neither, both before and after showing the reflections.

Finish with a class discussion about how $-f(x)$ and $f(-x)$ are different from $f(x)$ and from each other. Encourage students to use both algebraic and geometric reasoning.

Stretching and Shrinking Functions

When you analyze a new function, it's easier to understand the function's behavior and graph if you can recognize it as a transformed version of a function you already know. In this activity you'll explore function transformations that involve stretching or shrinking a parent function.

MULTIPLY THE VALUE OF THE FUNCTION

1. Open **Stretching Functions.gsp.** You will see the function graph $y = f(x)$. The function itself is not important to this investigation. It's just a curve with a distinctive shape. Point P is attached to the x-axis. The measurement x_p is the x-coordinate of P.

In this section you will transform the function by multiplying the value of the function by a constant.

2. Choose **Measure | Calculate** and calculate $f(x_p)$.

Enter function *f* by clicking the function definition on the screen and enter the value of x_p by clicking it on the screen.

3. Select in order x_p and $f(x_p)$. Choose **Graph | Plot As (x, y)**. A new point appears. Label it A.

Q1 Drag point P along the x-axis. What is the path of point A?

Now you will see what happens graphically when you multiply the value of the function by a constant.

4. Using the Sketchpad Calculator again, compute the value $2 \cdot f(x_p)$. Then plot point B with coordinates $(x_p, 2 \cdot f(x_p))$.

Q2 Where is point B in relation to point A? Is this relationship the same for all values of x? Test this by dragging point P.

Q3 Describe the path of B as you drag point P.

5. Select B and choose **Display | Trace Plotted Point**. Drag P again to trace the path of B.

Q4 The curve you have just traced is $y = 2 \cdot f(x)$. Describe it as a transformation of $y = f(x)$.

To erase the old traces, press the *Erase Traces* button at the bottom of the screen.

6. Double-click the calculation $2 \cdot f(x_p)$. Change it to $0.5 \cdot f(x_p)$. Drag point P and observe the path of B again.

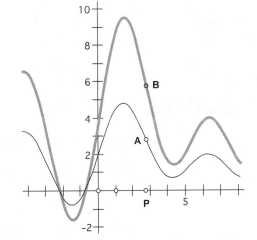

7. Show slider *a*. Change the calculation again, to make it $a \cdot f(x_p)$. Drag the slider and observe how point *B* behaves for different values of *a*.

To construct the locus, select both points and choose **Construct | Locus.**

8. To avoid the need to re-create traces every time you change the multiplier *a*, construct the locus of point *B* as *P* moves along the *x*-axis.

Q5 Use the slider to try various positive constants, some greater than 1 and some less than 1. Describe the transformation's effect on the graph when $a > 1$ and when $0 < a < 1$.

Q6 Try negative constants. Describe the transformation's effect on the graph when $a < -1$ and when $-1 < a < 0$. What happens when $a = 1$? When $a = 0$? When $a = -1$?

Q7 Double-click the original function to change its definition to a function of your own choice. Then try various values of *a*, and describe your results.

DIVIDE THE ARGUMENT OF THE FUNCTION

You just saw what happens when you multiply the function by a constant. In this section you will divide the argument (*x*) by a constant. You will take a shortcut by defining the transformed function directly, rather than transforming a point and constructing its locus.

9. Go to page 2. This page is similar to page 1, but the slider is labeled *b*. Set the value of *b* to 1.50.

To create the function, choose **Graph | Plot New Function.** To enter function *f* or value *b* into the new function, click the object in the sketch.

10. Create the transformed function in one step by graphing $g(x) = f(x/b)$.

Q8 Drag slider *b* to divide the argument by various constants. Try several different values, both positive and negative. Describe the effect on the transformed graph when $b > 1$, when $0 < b < 1$, when $b < -1$ and when $-1 < b < 0$.

Q9 Describe the shape of the transformed graph when $b = 1$, when $b = 0$, and when $b = -1$.

SUMMARY

11. Open page 3. The two sliders control the values *a* and *b*.

12. Define the function $g(x) = a \cdot f(x/b)$.

13. Drag each slider in turn and observe its effect.

Q10 Why do you think the directions asked you to *multiply* the value of the function by *a*, but to *divide* the argument by *b*?

Q11 Describe the graph of $g(x)$ as a transformation of $f(x)$.

Stretching and Shrinking Functions

Objective: Students stretch and shrink function graphs vertically and horizontally by multiplying the function value and dividing the function argument by constants.

Student Audience: Algebra 2

Prerequisites: Students must be familiar with function notation and graphing.

Sketchpad Level: Intermediate. Most of the work involves editing calculations and function definitions.

Activity Time: 20–30 minutes

Setting: Paired/Individual Activity (use **Stretching Functions.gsp**) or Whole-Class Presentation (use **Stretching Functions Present.gsp**)

Related Activity: Function Transformation Game

MULTIPLY THE VALUE OF THE FUNCTION

Q1 The path of A is $y = f(x)$, the function plot.

Q2 Point B is twice as far from the x-axis as A. This relationship holds no matter where point P is.

Q3 Since B is always twice as far from the x-axis as A, and A is always on the function graph, the path of B is a vertically stretched version of the path of A.

Q4 The graph of $y = f(x) + 4$ is the graph of $y = f(x)$ stretched in such a way that the positive values are twice as high and the negative values are twice as low.

Q5 For any function $f(x)$ and any constant $a > 1$, the graph of $y = a \cdot f(x)$ is the same as that of $y = f(x)$ stretched vertically by a factor of a. If $0 < a < 1$, the transformed graph shrinks toward the x-axis.

Q6 Negative values of a flip the graph across the x-axis. When $a < -1$, the flipped graph also stretches vertically. When $-1 < a < 0$, the flipped graph shrinks vertically. When $a = 1$, the transformed graph is identical to the original. When $a = 0$, the transformed graph is flat, identical to the x-axis (a constant function with a value of zero). When $a = -1$, the transformed graph is the reflection of the original across the x-axis.

Q7 Answers will vary, because students will choose different functions. But no matter what functions they choose, they should see the same behavior that's described for the previous question.

DIVIDE THE ARGUMENT OF THE FUNCTION

Q8 When $b > 1$, the graph stretches horizontally, so that every point on it is farther from the y-axis. When $0 < b < 1$, the graph shrinks horizontally, so that every point on it is farther from the y-axis. When $b < -1$, the graph reflects across the y-axis and stretches horizontally. When $-1 < b < 0$, the graph reflects across the y-axis and shrinks horizontally.

Q9 When $b = 1$, the transformed graph is identical to the original. When $b = 0$, the transformed graph does not exist, because division by zero is undefined. When $b = -1$, the transformed graph is a reflection across the y-axis.

SUMMARY

Q10 When $a > 1$, multiplying the value of the function results in stretching vertically, because the value of the function becomes greater. When $b > 1$, dividing the argument of the function means that a greater value of the argument must be used for the same result, so this results in a horizontal stretch. (This phrasing assumes positive values, but the same principle applies if the value of the function or argument is negative.) By using multiplication in one case and division in the other, both operations result in similar behavior: a stretch when the parameter is greater than 1 and a shrink when it's between 0 and 1.

Q11 For any function $f(x)$ and constants a and b, the graph of $y = a \cdot f(x/b)$ is the same as the graph of $y = f(x)$ transformed vertically by a stretch or shrink by a factor of a, and transformed horizontally by a stretch or shrink by a factor of b. If either a or b is negative, the graph is also reflected across the x- or y-axis.

Stretching and Shrinking Functions

Presenter Notes

In this presentation you will demonstrate stretches and shrinks of a graph in the form $y = a \cdot f(x/b)$ by showing the vertical and horizontal stretches and shrinks separately on pages 1 and 2, and then combining them on page 3.

1. Open **Stretching Functions Present.gsp.** Drag point P to show that it controls the measurement x_p. Drag the slider to show how it changes parameter a.

2. Press *Show A.* This point has coordinates $(x_p, f(x_p))$, so it must fall on the graph. Drag the slider so that $a = 2$.

Q1 What will happen if you multiply the function's value by a? If you plot the point $(x_p, a \cdot f(x_p))$, where will it fall? (It will be twice as far above the x-axis as point A.)

3. Press *Show B.* Drag P to show that B is always twice as far from the axis as A.

Q2 What will be the path of point B as you drag P left and right? (Its path is a vertical stretch of the graph of $y = f(x)$ by a factor of 2.)

4. Select point B and choose **Display | Trace Plotted Point.** Drag point P to trace the path.

Q3 What is the equation of this curve? ($y = 2f(x)$. Show it with other values of a by dragging slider a to a new value and then dragging P again.)

5. Go to page 2 and drag the slider so that $b = 2$.

Q4 Now you want to plot $(x_p, f(x_p/b))$. Should you evaluate the function at point P, using the argument x_p? (No, you need to divide x_p by 2 before evaluating.)

6. Press *Show A* to show the argument to use when evaluating the function. Press *Show B* to evaluate the function using this argument.

Q5 Is this the point you want to plot for the new function? (No, this gives the correct value to use, but students need to plot this value at the position of P on the x-axis.)

7. Press *Show C* to plot the value of the function at the position of P. Drag P back and forth, and have students observe how the graph is stretched horizontally.

8. Page 3 has sliders for both a and b. The combined graph $y = a \cdot f(x/b)$ is plotted. Try several different combinations for the two parameters, and challenge students to describe the stretch or shrink before you reveal it in the sketch.

9. Double-click the definition of $f(x)$ to edit it. Try several different definitions. Students need to understand that the principle of stretching or shrinking a function is independent of the specific choice of the parent function.

148 5: Algebraic Transformations

Exploring Algebra 2 with The Geometer's Sketchpad
© 2007 Key Curriculum Press

Transforming Odd and Even Functions

An *odd function* is one in which $f(-x) = -f(x)$ for all x in its domain.

An *even function* is one in which $f(-x) = f(x)$ for all x in its domain.

In this activity you'll transform a point using the above definitions and use the transformed point to test a variety of functions to see if they are odd or even.

TRANSFORM A POINT

1. In a new sketch, construct point A and measure its x- and y-coordinates.

Next transform these coordinates according to the definition of an odd function.

2. The transformed point must have an x-value opposite that of the original point, so calculate $-x_A$.

Q1 From the definition of an odd function, what calculation can you use to find the y-value that corresponds to an x-value of $-x_A$? Calculate this result.

To plot the transformed point, select the calculated x- and y-values in order and choose **Graph | Plot As (x, y)**.

3. Plot the transformed point and label the new point *Odd*.

Q2 From the definition of an even function, what calculations can you use to find the coordinates of a point that matches that rule?

4. Use the Calculator to compute these values. Plot the point and label it *Even*.

Q3 Describe the behavior of image points *Odd* and *Even* as you drag A.

TEST SOME FUNCTIONS

Select point A and the graph. Choose **Edit | Merge Point To Function Plot.**

5. Plot the function $f(x) = x^3 - 3x$. Merge point A to the new function plot.

6. Drag point A along the function plot and observe the transformed images.

Q4 How do the image points behave in relation to the function plot? Is the function even or odd? How can you tell from the image points?

Q5 Edit the function according to each problem below. Predict whether each will turn out to be even, odd, or neither. Then test by dragging A.

a. $f(x) = 5x$

b. $f(x) = x^3 + 2$

c. $f(x) = x^2 + 2$

d. $f(x) = x^4 - 3x^3$

e. $f(x) = \sin(x)$

f. $f(x) = \cos(x)$

Are there any functions that have odd symmetry about one point and even symmetry about a different point?

Q6 Some functions are neither odd nor even, but show even or odd symmetry about a point other than the origin. For instance, $f(x) = x^3 - 1$ shows odd symmetry about the point $(0, -1)$. Find two more such functions, one showing odd symmetry and one showing even. Identify the point or axis of symmetry, and edit the coordinate calculations to match the symmetry of the function.

Transforming Odd and Even Functions

Objective: Students transform the coordinates of a point according to the definitions of odd and even functions, and use the resulting images to explore the symmetry shown by the graphs of various specific functions.

Student Audience: Algebra 2

Prerequisites: None

Sketchpad Level: Challenging. Students should be familiar with the **Point** tool, with using the **Text** tool to label objects, with the Calculator, and with several commands from the Measure menu (**Abscissa** and **Ordinate**) and from the Graph menu (**Plot As (x, y)** and **Plot New Function**).

Activity Time: 25–35 minutes

Setting: Paired/Individual Activity (no sketch required) or Whole-Class Presentation (use **Transform Odd Even Present.gsp**)

TRANSFORM A POINT

1. Students must measure the *x*- and *y*-coordinates separately to use the results in calculations later.

Q1 In an odd function, the value of $f(-x)$ is the opposite of the value of $f(x)$, so the calculation should be $-y_A$.

3. The two calculations to plot are $(-x_A, -y_A)$.

Q2 In an even function, the value of $f(-x)$ is the same as the value of $f(x)$, so the values to plot are $(-x_A, y_A)$.

4. The two calculations to plot are $(-x_A, y_A)$.

Q3 Point *Even* is the reflection of *A* across the *y*-axis, and point *Odd* is its rotation by 180° about the origin. Students may also describe *Odd* as a double reflection across both *x*- and *y*-axes, as a dilation by −1 about the origin, or as a point reflection through the origin.

TEST SOME FUNCTIONS

Q4 As you drag *A* along the function plot, the *Odd* image traces out the opposite portion of the function plot. The *Even* image does not stay on the plot. This indicates that this particular function is odd, because the *Odd* image stays on it no matter where you drag *A*.

Q5 In each case students should test the function by dragging *A* along its plot. If the *Odd* image stays on the graph, the function is odd; if the *Even* image stays on the graph, the function is even; and if neither stays on it, the function is neither odd nor even. Here are the results:

  a. $f(x) = 5x$ is odd.

  b. $f(x) = x^3 + 2$ is neither.

  c. $f(x) = x^2 + 2$ is even.

  d. $f(x) = x^4 - 3x^3$ is neither.

  e. $f(x) = \sin(x)$ is odd.

  f. $f(x) = \cos(x)$ is even.

Q6 There are many possible answers.

Linear functions show point symmetry and are odd. (Any point on the graph can serve as the center.)

Second-degree (quadratic) polynomials are symmetric about a vertical axis and are even. (If the polynomial is expressed in the form $f(x) = a(x - h)^2 + k$, the axis of symmetry is the line $x = h$.)

Third-degree (cubic) polynomials are symmetric about a point and are odd. All these functions have a point of inflection, which is the point of symmetry.

Higher-degree polynomials generally exhibit neither kind of reflection symmetry. But some carefully chosen polynomials may show either kind of symmetry, depending on whether their highest-degree term has an even or odd exponent.

Absolute value functions have an axis of symmetry, and so exhibit even symmetry. Exponential and log functions exhibit neither odd nor even symmetry.

Sine and cosine graphs show both kinds of symmetry, requiring only a proper choice of point (for point symmetry) or vertical axis (for reflection symmetry). Because of their periodic property, these functions have infinitely many reflection points and reflection axes.

WHOLE-CLASS PRESENTATION

Use **Transform Odd Even Present.gsp** to present this activity to the class. Press the buttons and follow the directions on the screen to do the presentation.

6

Other Functions

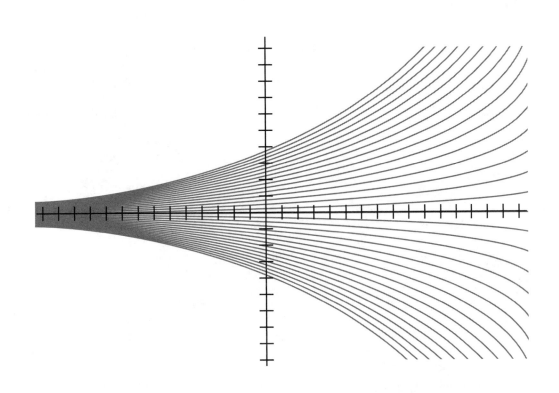

Absolute Value Functions

The absolute value of a number is how big the number is, regardless of whether it's positive or negative. Some examples should make this clearer.

The absolute value of -5 is 5,   *or* $|-5| = 5$.

The absolute value of 5 is 5,   *or* $|5| = 5$.

The absolute value of 0 is 0,   *or* $|0| = 0$.

As you can see, the absolute value of a number is always a positive number or zero.

But what happens when you graph an equation involving the absolute value function, such as $y = |2x - 4|$? In this activity you'll find out.

SKETCH AND INVESTIGATE

Choose **Graph | Plot New Function** to open the New Function calculator. Then type x and click OK.

1. In a new sketch, plot the equation $y = x$.

2. Think about what the graph of $y = |x|$ might look like. Plotting points by hand or discussing the question with classmates might help. Make a rough sketch of your guess on scratch paper.

Choose **Plot New Function** again. Then choose **abs** from the Functions pop-up menu, enter x, and click OK.

3. Plot the equation $y = |x|$. How does it compare with your prediction?

4. Repeat steps 1–3 with the equations $y = 2x - 4$ and $y = |2x - 4|$. Make sure to draw a prediction of what you think the second equation will look like before plotting it in Sketchpad.

Q1 Describe how the graphs of $y = 2x - 4$ and $y = |2x - 4|$ compare. Discuss their shapes, their ranges, and other features you notice.

Q2 Predict what the graph of $y = |2x| - 4$ will look like. After making your prediction, plot it. Then describe the differences between this graph and the other two.

A FAMILY OF ABSOLUTE VALUE GRAPHS

As you explore this family, keep in mind a related family: lines in point-slope form $y = m(x - h) + k$.

Now that you have an idea of what absolute values can do to the graphs of particular functions, you'll explore a *family* of graphs: $y = m|x - h| + k$.

5. Open **VGraph.gsp.** You'll see the graph of an equation in the form $y = m|x - h| + k$, and sliders for the parameters m, h, and k.

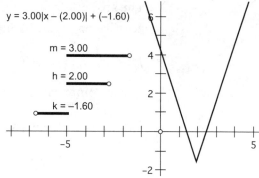

$y = 3.00|x - (2.00)| + (-1.60)$

$m = 3.00$

$h = 2.00$

$k = -1.60$

Q3 Adjust slider m and observe the effect this has on the V-graph. Try different values for the slope—large, small, positive, negative, and zero. Summarize the role m plays in the equation $y = m|x - h| + k$.

Q4 Changing m moves the entire V-graph except for one point. This point is called the *vertex*. Adjust the sliders for h and k. How does the location of the vertex relate to the values of h and k?

Q5 Write an equation in the form $y = m|x - h| + k$ for each of the V-graphs described below. Check each answer by adjusting the m, h, and k sliders.

   a. Vertex at $(-1, 2)$; contains the point $(0, 4)$

   b. Vertex at $(2, 3)$; contains the point $(4, 0)$

   c. Vertex at $(-3, -1)$; same shape as $y = 3|x - 2| + 5$

   d. x-intercepts at $(4, 0)$ and $(-4, 0)$; contains the point $(1, -2)$

   e. x-intercepts at $(6, 0)$ and $(-2, 0)$; contains the point $(5, 1)$

   f. Same vertex as $y = 3|x - 2| + 5$; contains the point $(0, 0)$

Q6 Not all graphs in this family have x-intercepts. How can you tell whether a function has x-intercepts just by looking at the function parameters?

EXPLORE MORE

To enter sin(x), choose **sin** from the Calculator's Functions pop-up menu. For this function's graph to appear properly, Sketchpad's angle units must be set to radians. Use **Edit | Preferences | Units** to check this setting.

Q7 Plot the following two pairs of equations:

   a. $y = x^2 - 1$ and $y = |x^2 - 1|$

   b. $y = \sin x$ and $y = |\sin x|$ (Don't worry if you're unfamiliar with the sin function.)

Write a short paragraph summarizing what happens when you plot a function and its absolute value.

Absolute Value Functions

Objective: Students graph the absolute value function and various transformations. In the process they review the point-slope form of linear functions and prepare for the vertex form of quadratic functions.

Student Audience: Algebra 1/Algebra 2

Prerequisites: Students should know the definition of the absolute value function, but they need no experience with the graph.

Sketchpad Level: Easy

Activity Time: 20–30 minutes

Setting: Paired/Individual Activity (use **VGraph.gsp**) or Whole-Class Presentation (use **VGraph Present.gsp**)

Related Activities: Parabolas in Vertex Form

In addition to familiarizing students with various transformations of the absolute value graph, there are two other reasons for doing this activity:

First, it reinforces students' understanding of the point-slope form of lines (with the role of point (h, k) becoming even more apparent).

Second, it prepares students for the vertex form of parabolas.

It's important for students to actually predict what the absolute value graphs will look like before they graph them. Just by plotting a few points—such as $(-2, 2)$, $(0, 0)$, and $(2, 2)$—they can gain insight into why these graphs appear as they do.

SKETCH AND INVESTIGATE

Q1 The graphs are identical to the right of $x = 2$. To the left of $x = 2$, the graphs are reflections on the x-axis. Another way of describing this is that the part of $y = 2x - 4$ that was below the x-axis has been flipped above the x-axis. The graph of $y = 2x - 4$ is a line whereas the graph of $y = |2x - 4|$ is shaped like a V. The range of $y = 2x - 4$ is all real numbers whereas the range of $y = |2x - 4|$ is all real numbers greater than or equal to zero.

Q2 This graph has the same parent function ($y = 2x - 4$), but a different vertex: the point $(0, -4)$. The purpose of this question is to spur students to think about how the numbers in the equation relate to the position of the vertex.

A FAMILY OF ABSOLUTE VALUE GRAPHS

Q3 The sign of m determines whether the V opens upward or downward or is flat. The graph of an equation with a positive m opens upward; with a negative m, it opens downward; and with $m = 0$, it is straight and horizontal. The greater the magnitude of m, the smaller the angle at the vertex. The slope of the right side of the V always equals m while the slope of the left side always equals $-m$.

Q4 The coordinates of the vertex are (h, k).

Q5 a. $y = 2|x + 1| + 2$ b. $y = -1.5|x - 2| + 3$
c. $y = 3|x + 3| - 1$ d. $y = (2/3)|x| - 8/3$
e. $y = -1|x - 2| + 4$ f. $y = -2.5|x - 2| + 5$

Q6 The graph does not have x-intercepts if parameters m and k have the same sign. For example, if k is positive, then the vertex is above the x-axis. If m is also positive, then the graph opens upward, so it can never reach down to the x-axis.

If m and k have opposite signs, then there are two x-intercepts. If $k = 0$, then the vertex itself is an x-intercept.

EXPLORE MORE

Q7 Consider the graphs of two equations: $y = f(x)$ and $y = |f(x)|$. Where the first graph is below the x-axis, it will be reflected across (above) the x-axis in the second graph. Where the first graph is above the x-axis, the second graph will be identical.

Absolute Value Functions

1. Open **VGraph Present.gsp.** Press *Show y = x.* After a short pause, press *Show y = | x |.*

Q1 It looks like the graphs coincide on the right side. Geometrically, what is their relationship on the left? (They are reflections on the *x*-axis.)

2. To verify this, construct a point on the left side of either graph. Double-click the *x*-axis to mark it as a mirror. Select the point and choose **Transform | Reflect.** Drag the first point to show that the image follows the other graph.

> There is also a reflection on the *y*-axis. If anyone mentions that, acknowledge it and verify that reflection too.

3. On page 2, press *Show y = mx + b.* Drag the two sliders to show how they control the graph.

Q2 Ask, "What will the graph of $y = | mx + b |$ look like?" Discuss predictions and show the graph.

Q3 Give *m* a positive value for this question. What is the slope on the right part of the absolute value graph? (*m*) What is the slope on the left side? ($-m$)

4. The slope on the right side should be clear because of the coinciding graphs. To check the slope on the left side, construct a line segment with both endpoints on the left side of the absolute value graph. Select the segment and choose **Measure | Slope.**

5. On page 3 is the graph of $y = m | x |$, with $m = 1$ to begin.

> There is a button that will show *m* as a fraction, reinforcing the slope concept.

Q4 Before moving the slider, ask for predictions. Then vary *m* and ask for observations as you make *m* bigger, smaller, negative, and zero. (As with multiplying any other function by a constant, *m* defines the ratio for a vertical stretch. In this case an *m* with larger magnitude makes the vertex angle smaller. If *m* is positive, it opens upward; if negative, downward. When $m = 0$, the graph coincides with the *x*-axis.)

6. Page 4 contains three sliders and the graph of $y = m | x - h | + k$.

Q5 What are the effects of changing the parameters? Parameters *h* and *k* define horizontal and vertical translation. Parameter *m* has the same effect as before.

Q6 Identify the vertex. What are the coordinates of this point? (*h, k*) To emphasize this, press *Reset,* and then drag the *h* and *k* sliders one at a time. Give students the coordinates of the vertex and one other point, and challenge them to derive corresponding settings for *m, h,* and *k*.

7. Page 5 has the graph of a general function along with its absolute value. Edit the definition of $f(x)$ and have students predict the shape of its absolute value. Challenge them with functions they have never seen. Given the root function graph, they should still be able to predict the shape of the absolute value.

Exponential Functions

There is a connection between population growth, radioactive decay, musical scales, and compound interest. They seem to have little in common, but you can model any of them using an exponential function.

An exponential function has the general form $f(x) = ab^x$, where $a \neq 0$, $b > 0$, and $b \neq 1$.

PROPERTIES OF THE GRAPH

Before using an exponential function to model a real-world problem, take some time to familiarize yourself with the graph.

To create a parameter, choose **Graph | New Parameter.**

1. In a new sketch, create parameters a and b.

2. Graph the function $f(x) = a \cdot b^x$ by choosing **Graph | Plot New Function.** Click the parameters in the sketch to enter them into the function.

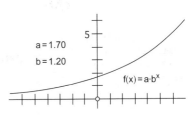

3. The graph is plotted on the screen. Change the values of the parameters, and observe the resulting changes in the graph. Try several values for each parameter.

To change the value of a parameter, either double-click it or select it and press the + or − key. (To change the size of the steps for the + or − keys, select the parameter and choose **Edit | Properties | Parameter.** Change the Keyboard Adjustments value to 0.1 unit.)

Q1 What are the x- and y-intercepts of the graph? Explain how the intercepts are related to parameters a and b.

Q2 The value of $f(x)$ tends to get close to zero either on the left side or on the right. What parameter values determine which side it is?

Q3 In the general form of the exponential function, there are three constraints ($a \neq 0$, $b > 0$, and $b \neq 1$). Explain the reason for each of these constraints.

Next you'll investigate how the function behaves by comparing the coordinates of two points on the graph.

4. Change the parameters so that $a = 2.00$ and $b = 1.30$.

5. Construct two points on the function graph. Label them P and Q.

6. Select both points and choose **Measure | Abscissa (x).** Select the points again and choose **Measure | Ordinate (y).**

7. Calculate the values $x_Q - x_P$ and y_Q/y_P.

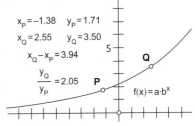

You can change the scale of the axes to give you more precise control over the positions of the points.

Q4 Drag point Q one unit to the right of P, so that the difference $x_Q - x_P$ is as close to 1.00 as you can make it. What is the value of the ratio y_Q/y_P? Drag point P to a different position on the graph, and again drag Q so it's one unit to the right. What is the value of the ratio y_Q/y_P? Why do you get this result?

Q5 Now use a difference other than 1. Drag the points so that $x_Q - x_P$ is approximately 2.5. What is the ratio? Drag them to another position, but with the x difference still equal to 2.5. What is the ratio now? Explain.

DOUBLING PERIOD AND HALF-LIFE

Exponential functions can be used to solve a number of real-life problems. First change the function so that it shows the value of $100 invested at an effective annual yield of 6%.

Effective annual yield is not the same thing as interest rate. That's another topic.

8. Edit parameters a and b so that the function has the definition $f(x) = 100(1.06)^x$. This shows the value of $100 invested at an effective annual yield of 6%. (The x variable is in years, and y is in dollars.)

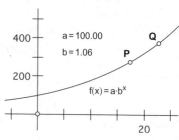

Setting the grid to rectangular allows you to adjust each axis independently of the other.

9. At first you can't see the graph because the y-axis doesn't go up to 100. To adjust the axes, choose **Graph | Grid Form | Rectangular Grid.** Then drag tick mark numbers on each axis so that you can see the results for the first 25 years.

To see more decimal places in a parameter, select the parameter and choose **Edit | Properties | Value.** Change the Precision setting.

Q6 How long will it take to double your money? Drag the points so that the ratio is 2.00. What is the difference in their x-coordinates? This number is called the *doubling period.*

An exponential function can also be used to model the decay of radioactive cesium.

10. To model the decay of 80 g of cesium, change the function definition to $f(x) = 80(0.977)^x$. Adjust the axes appropriately. The value of x is still in years, but the value of y is in grams.

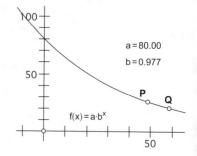

Q7 If you start with 80 g, you will have less cesium every year. How long would it take to lose half of it? Explain how you found the answer. This number is called the *half-life* of cesium.

Q8 Although cesium decays, as opposed to growing, you can still calculate its doubling period. Drag the two points until you find a position where the ratio is 2.00. What is the difference in the x-coordinates? Explain how this verifies your answer to Q7.

Exploring Algebra 2 with The Geometer's Sketchpad
© 2007 Key Curriculum Press

Exponential Functions

Objective: Students graph exponential functions and examine their properties. They use exponential functions to model compound interest and radioactive decay.

Student Audience: Algebra 2/Precalculus

Prerequisites: Students must first understand how to work with expressions with exponents.

Sketchpad Level: Intermediate. Students create their own sketches using the graphing tools.

Activity Time: 30–40 minutes

Setting: Paired/Individual Activity (no sketch required) or Whole-Class Presentation (use **Exponential Present.gsp**)

PROPERTIES OF THE GRAPH

Q1 Both a and b are nonzero. Therefore, ab^x is also nonzero for any real x, so y cannot be zero, and there is no x-intercept.

For the y-intercept, substitute zero for x in the equation $y = ab^x$.

$$y = ab^0 = a$$

The y-intercept is a.

Q2 For $b > 1$, $f(x)$ tends to zero on the left. For $0 < b < 1$, $f(x)$ tends to zero on the right. The value of a has no influence on this property.

Q3 If a were equal to zero, the function would be the constant function $f(x) = 0$.

If b were equal to zero, the function would be zero for all positive x, and it would be undefined for all other values of x.

If b were less than zero, b^x would not be continuously defined over all real exponents x.

If b were equal to one, this would be another constant function, $f(x) = a$.

Q4 Limited resolution may prevent students from making the difference exactly 1.00. It's sufficient for them to make it as close to that value as they can. The ratio $y_Q/y_P = 1.30$. This is the same as parameter b.

$$\frac{y_Q}{y_P} = \frac{f(x_Q)}{f(x_P)} = \frac{ab^{x_Q}}{ab^{x_P}} = b^{x_Q - x_P} = b^1$$

Here it does not matter where on the graph points P and Q are, so long as $x_Q - x_P = 1$.

Q5 If $x_Q - x_P = 2.5$, then $y_Q/y_P = 1.30^{2.5} \approx 1.9$. This follows from the same reasoning as in the previous answer.

$$\frac{y_Q}{y_P} = \frac{f(x_Q)}{f(x_P)} = \frac{ab^{x_Q}}{ab^{x_P}} = b^{x_Q - x_P} = b^{2.5}$$

DOUBLING PERIOD AND HALF-LIFE

8. There may be some confusion regarding the term *effective annual yield*. The actual rate is about 5.8%, but since it is compounded continuously, at the end of each year, the investment will be worth 6% more than it was at the beginning of the year. Even if students do not yet grasp the concept, they can continue with the given function definition.

Q6 The doubling period is about 11.90 years.

Q7 To find the half-life, students should arrange the points so that $y_Q/y_P = 0.50$. When that happens, $x_Q - x_P \approx 30$, so cesium has a half-life of about 30 years. In this case changes in the scales of the axes can cause quite a lot of variation in the answers.

Q8 To find a doubling period, with the ratio $y_Q/y_P = 2.00$, point Q would have to be to the left of P. In that case, $x_Q - x_P \approx -30$.

This answer fits with the half-life answer. It stands to reason that if it takes 30 years for half of the cesium to decay, then 30 years ago, there was two times as much. This explains the negative doubling period.

1. Begin by showing the general form of this exponential function:

$$f(x) = ab^x, \quad \text{where } a \neq 0, b > 0, \text{ and } b \neq 1$$

Press *Show Slider Controls* to reveal buttons that you can use to set the parameters to various precise values.

2. Open **Exponential Present.gsp.** This is a graph of the function with sliders controlling the values of *a* and *b*. Drag each slider in turn so that students can see the effects on the graph.

Q1 Ask students for the *x*- and *y*-intercepts. (There is no *x*-intercept, and the *y*-intercept is equal to *a*.) Challenge them to verify these facts by alternately setting *y* and *x* to zero in the equation $y = ab^x$.

Q2 Sometimes the function approaches zero on the left side, and sometimes on the right. What determines which side it is? (It's on the left when $b > 1$ and on the right when $0 < b < 1$.)

Q3 In the general form of the function, there are three restrictions on the parameters *a* and *b*. Why? Show students what happens when $a = 0$, $b < 0$, $b = 0$, or $b = 1$.

3. On page 2 there are two points on the graph, *P* and *Q*. You can drag *P* freely, but point *Q* is a fixed distance to the right of point *P*. That distance is determined by the slider labeled Δ. At the bottom of the screen are measurements showing the difference of the *x*-coordinates and the ratio of the *y*-coordinates.

4. Drag point *P* to show that the ratio of the *y*-coordinates remains constant when the difference in the *x*-coordinates (Δ) is constant.

On pages 3 and 4, the *a* and *b* sliders have been replaced with parameters in order to make it easier to enter precise values.

Q4 What will the ratio be when $\Delta = 1.00$? (It should equal *b*.) Challenge students to predict this before showing it. Then have them prove that $\dfrac{f(x_P + 1)}{f(x_P)} = b$.

5. Page 3 has a graph showing the growth of an investment of $100 with an effective annual yield of 6%.

Pages 3 and 4 have rectangular grids. If students have not used that feature yet, this would be a good opportunity to show them the advantages of using different scales on the axes.

Q5 With this investment, how long would it take to double your money? If the money is doubled between *P* and *Q*, then the ratio of their *y*-coordinates will be 2. Drag the Δ slider until the ratio is 2.00. (The doubling period is the difference in the *x*-coordinates, about 11.9 years.)

6. The graph on page 4 shows the radioactive decay of 80 g of cesium. The *x*-scale is in years.

Q6 What is the half-life of cesium? This question is similar to the previous one. Give students time to figure out that they need to adjust the difference in the *x*-coordinates so that the ratio is 0.50. (The result is a half-life of about 30 years.)

Logarithmic Functions

Many occurrences in our natural world can be modeled using logarithmic functions, including the strength of earthquakes, the intensity of sound, or the concentration of hydronium ions in a solution. In this activity you'll explore the relationship between exponential and logarithmic functions, and determine how to write the formula for a logarithmic function that's the inverse of a particular exponential function.

GRAPH INVERSE EXPONENTIAL FUNCTIONS

Logarithms are related to exponents, so start by graphing an exponential function and finding the inverse graph.

1. Open **Logarithmic Functions.gsp.** Press the *Show Exponential Function* button to see the exponential function $y = 2^x$ along with its graph.

With the points selected, choose **Measure | Coordinates** to find their ordered pairs. With the coordinates selected, choose **Graph | Tabulate** to place the coordinates in a table.

2. Press the *Show Points* button to show seven points on the curve. Measure their coordinates, and put the resulting ordered pairs into a table.

Q1 Notice that some of the *x*-values are negative. Does this mean that the resulting values of the function are negative? Explain why this is true or not true.

Next, interchange the *x*- and *y*-values by reflecting the points over the line $y = x$.

3. Press the *Show y=x Line* button. With the line selected, choose **Transform | Mark Mirror.** The line flashes briefly to indicate that it is marked as the mirror.

To show the labels of the selected points, choose **Display | Show Labels.**

4. Press *Show Points* again to select the seven points in order, and choose **Transform | Reflect.** The seven points are reflected. Show their labels.

5. Measure the coordinates of the reflected points and tabulate the results. Align the two tables in order to see the original and reflected points next to each other.

Q2 What do you notice about the coordinates of each pair of points?

Q3 Will any of the *x*-coordinates of the reflected points be negative? Explain.

Q4 Why is the line $y = x$ called the *axis of symmetry* for a function and its inverse?

Next reflect the entire graph over the line $y = x$.

Use the **Point** tool to construct the new point.

6. Construct a point on the original graph, and reflect it over the line $y = x$. Drag your new point and observe the behavior of the reflected image.

The reflected graph is the graph of the inverse of the original function.

7. To create the entire reflected graph, select the point on the graph and its reflected image, and choose **Construct | Locus.** Change the color of the locus, and make it dashed.

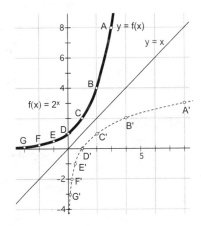

You'll use page 2 to graph $y = 10^x$, reflect it to show its inverse, and compare the inverse to the graph of $y = \log x$.

8. On page 2 construct the graph of $y = 10^x$ by choosing **Graph | Plot New Function** and entering 10^x into the Calculator.

9. Construct a point on the graph and reflect it across the graph of $y = x$.

To turn tracing on or off, select the point and choose **Display | Trace Point.**

10. Turn on tracing for the reflected point, and drag the point on the graph to observe the shape of the inverse function.

11. Construct the graph of $y = \log x$.

Q5 What do you observe about the graph of the log function and the reflected image of the exponential graph? What conclusion can you draw?

On page 3 you'll graph the exponential function $f(x) = k^x$ and use different values for k to find a general formula for the logarithmic function that's the inverse of $f(x)$.

Choose **Graph | New Parameter** *to create the new parameter, and set its label to* k *and its value to 2 in the dialog box that appears.*

12. Create a parameter k, set its value to 2, and use it to construct the graph of $y = k^x$. To enter k into the function definition, click its value in the sketch.

13. Construct a point on the graph, reflect it across $y = x$, and construct the locus.

This locus is the graph of the inverse function. Express this inverse as a logarithmic function by stretching or shrinking the parent logarithmic function $y = \log x$.

14. Using the values of a and b, plot the logarithmic function $y = a\log(x/b)$. Adjust the sliders so that your newly plotted function matches the inverse of $y = k^x$.

Q6 What values of a and b made the graphs match?

To add a row to a table, making the current values permanent, double-click the table.

15. Record the values of k, a, and b in a table. Then change the value of k to 5, match the graphs again, and add a new row of data to the table. Continue adding new data to the table for the following values of k: 10, 100, and 1000.

Q7 What pattern can you find to relate the value of k to the values of a and b?

Q8 Use this pattern to predict the values of a and b needed when $k = 10{,}000$. Test your prediction by gathering another row of data for your table. Then predict the values needed when $k = 0.1$, and test your prediction.

Q9 Use your results to write a formula for the inverse of $f(x) = k^x$.

EXPLORE MORE

Q10 Use algebraic manipulation to explain why your formula from Q9 must be true.

Q11 A general exponential function can be written as $f(x) = a \cdot 10^{(x-h)/b} + k$. Write the corresponding inverse function in terms of a, b, h, and k.

Exploring Algebra 2 with The Geometer's Sketchpad
© 2007 Key Curriculum Press

Logarithmic Functions

Objective: Students explore the graphical relationship between exponential and logarithmic functions and determine how to write the formula for a logarithmic function that's the inverse of a particular exponential function.

Student Audience: Algebra 2

Prerequisites: Students should be familiar with exponential functions and with the properties of inverse functions.

Sketchpad Level: Intermediate/Challenging. This activity involves a fair amount of graphing and construction. Students should be able to repeat later in the activity a set of steps they used earlier.

Activity Time: 40–50 minutes. You can do steps 9–12 as a presentation to save time.

Setting: Paired/Individual Activity (use **Logarithmic Functions.gsp**) or Whole-Class Presentation (use **Logarithmic Functions Present.gsp**)

Related Activities: Exponential Functions

GRAPH INVERSE EXPONENTIAL FUNCTIONS

On page 1 students graph $y = 2^x$ to begin with an exponential function that has points whose coordinates they can verify in their heads. On page 2 they use a base of 10 to make it easy to see that 10^x and $\log x$ are inverse functions. On page 3 they use function transformations to find the inverse of a more general exponential function, and end up with a conjecture as to how to write the formula for such a function. This portion of the activity is useful either before introducing the change-of-base property of logarithms (to motivate that method) or afterward (to review the method).

2. When students press *Show Points,* the seven points are shown and selected. Because the points are selected in order, it's easy for students to measure and then tabulate the coordinates. Be sure they measure the coordinates using **Measure | Coordinates** rather than measuring the x- and y-values separately. If they deselect all objects at any point in the process, they can press *Show Points* to select objects again in order.

Q1 A negative value of x in the exponential function means that the exponent is negative. A positive base raised to a negative exponent results in a positive number.

Q2 Within each pair of points (A and A', B and B', and so forth), the x- and y-values are interchanged.

Q3 None of the x-coordinates of the reflected points can be negative, because none of the y-coordinates of the original function were negative.

Q4 The line $y = x$ is called the *axis of symmetry* because it lies precisely between the two graphs, reflecting each onto the other.

Q5 The graph of the log function coincides with the reflected image of the exponential graph, so the log function must be the inverse of the original exponential function.

13. This instruction is brief and does not give precise details concerning commands. If students have trouble with it, have them review what they did in steps 7–8.

Q6 The two graphs match when a is approximately 3.32 and $b = 1$. This value of $a = 3.32$ is approximately $1/\log 2$, but students don't have enough information yet to reach this conclusion.

Q7 When k is 10^1, $a = 1$; when k is 10^2, $a = 1/2$; when k is 10^3, $a = 1/3$. It appears that a must be $1/\log k$.

Q8 Because $10{,}000 = 10^4$, the value of a must be $1/4$. Because $0.1 = 10^{-1}$, the value of a must be $1/-1 = -1$. Students should verify these predictions by testing them in the sketch.

Q9 The inverse of $f(x) = k^x$ is $f^{-1}(x) = \log x / \log k$.

EXPLORE MORE

Q10 Define $d = \log k$. Rewrite k as 10^d. Then:

$$y = k^x = (10^d)^x = 10^{dx}$$

so $\log y = dx$

and $x = \dfrac{\log y}{d} = \dfrac{\log y}{\log k}$

Q11 The inverse function is $f^{-1}(x) = b\log((x - k)/a) + h$.

Logarithmic Functions

In this presentation students will find the log function that's the inverse of $f(x) = k^x$.

Remind students that the ordered pairs of an inverse function are in the opposite order.

1. Open **Logarithmic Functions Present.gsp.** Press *Show Exponential Function.* The graph of $f(x) = 2^x$ appears. Press *Show Points* and then *Show Coordinates* to show seven points on the graph, along with their coordinates.

Q1 On the inverse function, what are the coordinates of the point corresponding to *A* on the original function? (8.00, 3.00) Ask different students to give coordinates for points on the inverse corresponding to each of the other points.

Q2 What geometric transformation do you know that switches the *x*- and *y*-values? (reflection across the line $y = x$)

2. Use the next three buttons to reflect the points across $y = x$ and show the coordinates. The two tables confirm the answers students gave for Q1.

3. To see the entire graph of the inverse function, press *Show Reflected Graph.*

Now compare an exponential function with a logarithmic function, and demonstrate that they are inverses by verifying that each is a reflection of the other across $y = x$.

4. On page 2, use the top two buttons to show the exponential function $f(x) = 10^x$ and the logarithmic function $g(x) = \log x$.

5. Use the next two buttons to show $y = x$, a point on $f(x)$, and its reflection.

6. Use the Animate button to animate the point, trace the reflection, and verify that the reflection (and therefore the inverse) of $f(x) = 10^x$ really is $g(x) = \log x$.

Now graph an exponential function with an adjustable base, and graph a stretchable logarithmic function. Match the logarithmic function to the inverse of the exponential function, and find a formula for the inverse of the exponential function.

7. On page 3, press the first three buttons to show the function $f(x) = k^x$ and its inverse. Use the green buttons to switch *k* from 2 to 5 and back to 2 again.

8. Show the adjustable logarithmic function, and drag sliders *a* and *b* to make the red logarithmic function match the blue inverse function.

9. Show the table of values for *k*, *a*, and *b*. Double-click the table to make the current values permanent.

10. Set *k* to the values 5, 10, 100, and 1000. For each value of *k*, drag sliders *a* and *b* to match the inverse, and record the values permanently in the table.

Q3 Ask, "How can you determine the values of *a* and *b* from *k*?" (The value of *b* is always 1, and the value of *a* is 1/log *k*.)

11. Test this conclusion by using the remaining green buttons to adjust *k*, predicting the required values of *a* and *b*, and testing the predictions by dragging *a* and *b*.

Exploring Algebra 2 with The Geometer's Sketchpad
© 2007 Key Curriculum Press

Square Root Functions

An important formula in geometry is the formula for the area of a square. In this activity you'll start with this formula to gain an understanding of the square root function and to explore transformations of this function.

INVESTIGATE

You'll use a model for the area of a square to begin your investigation.

1. Open **Square Root Fns.gsp.** This page shows, in two different ways, the relationship between the side and the area of a square.

2. Use the slider below the top square to drag *s* left and right.

Q1 Drag *s* so it's exactly 2.00. What is *a*? Make *s* exactly 5.00. What is *a* now? What happens to *a* when you move *s* to the negative side? Without using specific numbers, describe the relationship between *s* and *a*.

Q2 In this model, does *s* depend on *a*, or does *a* depend on *s*? Which of these two is the independent variable? How can you tell?

Q3 Write a formula for *a* in terms of *s*.

3. To plot a data point, select the two measurements in order (independent variable first, then dependent variable). Choose **Graph | Plot As (x, y)**, and then choose **Display | Trace Plotted Point.** Label the point *P*.

4. To see the entire graph, vary the variable by dragging *s* left and right.

In the bottom square the roles of the variables are reversed.

Q4 Drag *a* so it's exactly 9.00. What is *s*? Make *a* exactly 32.00. What is *s* now? If *a* were 76.00, between what two positive integers would *s* be? Explain.

5. In this case, what are the independent and dependent variables? Plot this new point as in step 3. Label the point *Q*.

Q5 Drag *a* to the left and to the right. How does *Q* behave? What happens when you drag *a* to the left of zero? Explain why this happens.

To plot a function, make sure nothing is selected and choose **Graph | Plot New Function.** Square root appears as **sqrt** on the function pop-up menu.

6. Plot the functions $f(x) = x^2$ and $g(x) = \sqrt{x}$.

Q6 How do these graphs compare to the existing traces?

So far you have plotted inverse functions algebraically, by changing which variable is the independent variable and which is the dependent. On pages 2 and 3 you'll look at the process geometrically.

Square Root Functions
continued

7. On page 2 are point A and the measurements of its x- and y-coordinates. To form an inverse relation, you must interchange these coordinates, using y for x and x for y. Interchange the coordinates of A by plotting the point (y_A, x_A).

8. Turn on tracing for both points, and drag point A around the coordinate plane. Try to make some interesting shapes.

Q7 What kind of symmetry do your patterns show?

To erase all traces, choose **Display | Erase Traces.**

9. Erase the traces and drag A again. This time try to keep A and the plotted point together. This process will reveal the line of symmetry.

Q8 What is the equation of the line of symmetry?

Next you'll use the line of symmetry to construct the inverse of $y = x^2$.

To mark a mirror, select the line and choose **Transform | Mark Mirror.**

10. Page 3 shows the graph of $f(x) = x^2$ and the line of symmetry $y = x$. Mark the $y = x$ line as the mirror. Use the **Point** tool to put a point on the graph. Then reflect the point across the line.

To trace the point, select it and choose **Display | Trace Point.**

11. Drag the point on the graph. Turn on tracing for the reflected point, and drag again.

Q9 How does the reflected image point behave? Does it appear to trace out a function? (*Hint:* Try the vertical line test on it.)

Q10 Plot the function $g(x) = \sqrt{x}$. How does this function plot differ from the reflected image?

EXPLORE MORE

12. Page 4 shows the graph of $y = \sqrt{x}$ and the line of symmetry $y = x$. As before, put a point on the graph, reflect it across the line of symmetry, turn on tracing, and drag the point on the graph.

Q11 Plot the function $g(x) = x^2$. This time, how does the function plot differ from the reflected image?

Q12 Is either of the two reflections you constructed a function graph? Can you say that $f(x) = x^2$ and $g(x) = \sqrt{x}$ are inverse functions? If not, what restrictions must you specify in order to have inverse functions?

Square Root Functions

Objective: Students explore the square root function by generating data: First they vary the side length of a square and observe its area, and then they vary the area and observe the resulting side length. They investigate the relationship between these two functions using both algebraic and geometric methods. The activity ends by considering whether the inverse of x^2 is a function, and spurs students to think about the conditions under which inverse relations are also inverse functions.

Student Audience: Algebra 2

Prerequisites: None

Sketchpad Level: Intermediate

Activity Time: 30–40 minutes

Setting: Paired/Individual Activity (use **Square Root Fns.gsp**) or Whole-Class Presentation (use **Square Root Fns Present.gsp**)

INVESTIGATE

Q1 When $s = 2$, $a = 4$. When $s = 5$, $a = 25$. When s is negative, a remains positive. As the value of s gets farther from zero, the value of a gets larger even faster.

Q2 In this model s is the value you change, so it's the independent variable, and a is the dependent variable.

Q3 The formula is $a = s^2$.

Q4 When $a = 9$, $s = 3$. When $a = 32$, $s = 5.66$. If a is 76, s must be between 8 and 9, because $8^2 = 64$, $9^2 = 81$, and 76 is between 64 and 81.

Q5 When you drag a to the right, Q follows a curving path in Quadrant I. When a is to the left of zero, point Q disappears, because the model does not allow a negative area.

Q6 The graphs of the two functions appear to match the traces that result from dragging the sliders.

Q7 The traces show reflection symmetry about a line that goes diagonally up and to the right.

Q8 The line of symmetry is $y = x$.

Q9 The reflected point traces out a full reflection of the x^2 function plot, in the shape of a parabola opening to the right. This trace does not define a function plot, because it fails the vertical line test.

Q10 The plot of $g(x)$ coincides with the trace in Quadrant I, where y is positive. However, the function plot stops at the origin and does not go below the x-axis, and so leaves out the lower branch of the parabola. The trace includes both the upper and lower branches of the parabola.

EXPLORE MORE

Q11 The reflected image of the square root function consists of the right-hand branch of a parabola. The plotted function $g(x) = x^2$ defines both branches of the parabola.

Q12 The reflection of $f(x) = x^2$ is not a function graph, because the inclusion of points like $(4, 2)$ and $(4, -2)$ makes it fail the vertical line test. The reflection of $y = \sqrt{x}$ is a function graph, consisting of the portion of $y = x^2$ for which $x \geq 0$.

If we restrict $f(x) = x^2$ to the domain $x \geq 0$, its inverse is a function: $g(x) = \sqrt{x}$. (This is called a *one-to-one function*.)

WHOLE-CLASS PRESENTATION

Use the Presenter Notes and **Square Root Fns Present.gsp** to present this activity to the whole class.

Use this presentation to explore the square root function, to investigate its relationship to $y = x^2$, and to stimulate students to think about inverse relations and inverse functions.

1. Open **Square Root Fns Present.gsp.** This page shows in two different ways the relationship between the side and the area of a square.

2. Drag slider s and have students observe the changing values of s and a. Then show the plotted point P and continue dragging the slider to trace out the graph.

Q1 How can you tell which is the independent variable? Write a formula for a in terms of s. ($a = s^2$)

3. Drag slider a for the lower square and have students observe the changing values. Show the plotted point Q and continue dragging to trace out the graph.

Q2 How can you tell which is the independent variable? Write a formula for s in terms of a. ($s = \sqrt{a}$)

Q3 Ask students to describe each graph individually, and then to compare the two. What is similar about them? How are they related to each other? What are the important differences?

4. Press *Show x^2* and *Show sqrt(x)*.

Q4 How do these graphs compare to the existing traces?

So far you have plotted inverse functions algebraically, by changing which variable is the independent variable and which is the dependent. On page 2 you'll look at the process geometrically.

5. Page 2 shows the graph of $y = x^2$ and the line $y = x$. There's a point on the graph of $y = x^2$. To reflect this point across $y = x$, click *Show Point Reflection.*

Q5 Drag the point on the graph. Turn on tracing for the reflected point and drag again. How does the reflected image point behave?

Q6 Click *Show Function Reflection* to see the entire function reflected. How is it related to the reflected image?

Q7 Is the reflection the plot of a function? Does it pass the vertical line test?

Q8 How does the reflection compare to the graph of the function $f(x) = \sqrt{x}$?

Q9 On page 3, compare the reflection of $f(x) = \sqrt{x}$ to the graph of $y = x^2$.

Q10 Discuss with students how to modify the situation so that you have two functions, each of which is the inverse of the other. What sort of restrictions would be necessary?

© 2007 Key Curriculum Press

Rational Functions

A rational number gets its name because it is the ratio of two integers. Rational functions are named for similar reasons. A rational function is the ratio of two polynomial functions.

If $f(x)$ is a rational function, then $f(x) = \frac{p(x)}{q(x)}$, where $p(x)$ and $q(x)$ are polynomial functions. The polynomials can be of any degree, but in this investigation they will have degree one or zero.

THE RECIPROCAL FUNCTION

One simple example of a rational function is the reciprocal function $1/x$. This will be the first function in your investigation.

1. In a new sketch, define the coordinate axes. Adjust the scale so that the x-axis fits your screen between about -10 and 10.

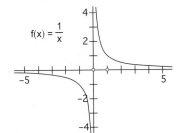

2. Choose **Graph | Plot New Function.** Define the new function $f(x)$:

$$f(x) = \frac{1}{x}$$

Q1 Does $f(x)$ satisfy the definition of a rational function? Explain.

Q2 For what values of x is $f(x)$ zero? For what values is it undefined?

The graph $y = f(x)$ is a hyperbola. Like all hyperbolas, it has two *asymptotes*, lines that the curve approaches at the extremes.

Q3 What are the equations of the asymptotes of this curve? Asymptotes are lines, so your answer should be the equations of two lines.

TRANSFORMATIONS OF THE RECIPROCAL FUNCTION

Use color to match each function with its graph. Select the function definition and its graph, and choose **Display | Color** to set a new color.

3. On the same grid, plot another rational function:

$$g(x) = \frac{4x - 10}{x - 3}$$

Q4 What are the equations of the asymptotes of the graph of $g(x)$?

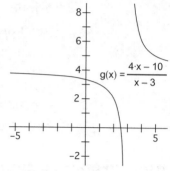

Here's how it works. The function $g(x)$ is a fraction. Divide the denominator into the numerator, and leave a remainder.

$$g(x) = \frac{4x - 10}{x - 3} = \frac{4(x - 3) + 2}{x - 3} = \frac{2}{x - 3} + 4$$

When you view the function this way, you can see $g(x)$ in terms of the reciprocal function. Since $f(x) = 1/x$, it follows that $g(x) = 2f(x - 3) + 4$. Expressed as a transformation, this stretches the parent function, $f(x)$, vertically by a ratio of 2, and translates it right 3 units and up 4 units.

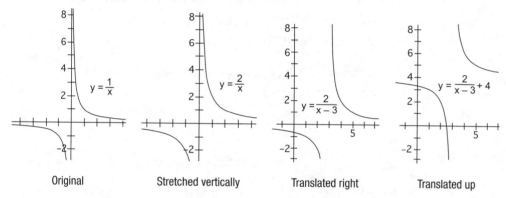

| Original | Stretched vertically | Translated right | Translated up |

Q5 How does this transformation help explain the positions of the asymptotes of $g(x)$?

You can verify this by plotting a third function as a transformation of $f(x)$. Its graph should fall right on top of the graph of $g(x)$.

4. Plot $h(x)$ with the following definition:

$$h(x) = 2f(x - 3) + 4$$

Q6 Below are three new definitions to try for $g(x)$. In each case, express the function in terms of $f(x)$, as in the example above. Find the asymptotes of each graph. Check your work by plotting $g(x)$ and $h(x)$ in the sketch.

a. $g(x) = \dfrac{-3x - 12}{x + 5}$ b. $g(x) = \dfrac{4x - 17}{4x - 16}$ c. $g(x) = \dfrac{15x + 103}{5x + 35}$

Q7 Given the two asymptotes, it is possible to find any number of different rational functions that fit them. Define two different functions having the asymptotes $x = 6$ and $y = -4$. Start by showing the functions as transformations of $f(x)$, and then express them as ratios of polynomials.

Exploring Algebra 2 with The Geometer's Sketchpad
© 2007 Key Curriculum Press

Rational Functions

Objective: Students define simple rational functions and show their graphs as transformations of the graph of the reciprocal function: $y = 1/x$.

Student Audience: Algebra 2

Prerequisites: This activity depends heavily on the concept of function transformation. Students should be comfortable with that before they begin.

Sketchpad Level: Intermediate. Students graph three functions, building the sketch from scratch.

Activity Time: 30–40 minutes

Setting: Paired/Individual Activity (no sketch required) or Whole-Class Presentation (use **Rational Functions Present.gsp**)

THE RECIPROCAL FUNCTION

Q1 Yes, this does fit the definition. Both 1 and x are polynomials. You may need to remind students that a polynomial may have only one term, and it does not necessarily have to have a variable.

Q2 The function $f(x)$ is not zero for any real x. It is undefined for $x = 0$.

Q3 The asymptotes are the x- and y-axes. Their equations are $x = 0$ and $y = 0$.

TRANSFORMATIONS OF THE RECIPROCAL FUNCTION

Q4 The asymptotes of $g(x)$ are the lines $x = 3$ and $y = 4$.

Q5 The parent function, $f(x) = 1/x$, has asymptotes at the x- and y-axes. The stretching transformation does not change either of these lines. The translations move the vertical asymptote right 3 units to $x = 3$ and the horizontal asymptote up 4 units to $y = 4$.

Q6

	Transformation Function	Asymptotes	
a.	$3f(x + 5) - 3$	$x = -5$	$y = -3$
b.	$-\frac{1}{4}f(x - 4) + 1$	$x = 4$	$y = 1$
c.	$-\frac{2}{5}f(x + 7) + 3$	$x = -7$	$y = 3$

Q7 Answers will vary. For any pair of asymptotes, $x = h$ and $y = k$, it is possible to write the function as a transformation in this form:

$$g(x) = a \cdot f(x - h) + k$$

This transformation stretches the graph vertically by a factor of a and translates it to new center point (h, k). In this instance $h = 6$ and $k = -4$. Since a, the ratio of the stretch, can be any real number, there is no limit to the number of unique rational functions that fit these asymptotes.

To express the function as a ratio of polynomials, substitute 6 for h and -4 for k, and use $f(x) = 1/x$. Then combine the parts into a single fraction.

$$g(x) = a \cdot f(x - 6) - 4 = \frac{a}{x - 6} - 4 = \frac{-4x + 24 + a}{x - 6}$$

Any function in this form will have the required asymptotes.

EXTENSION

Throughout this activity the stretching transformations are called *vertical stretches*. In fact, because of a special property of the reciprocal function, you could just as well call them *horizontal stretches*.

Again let $f(x) = 1/x$. To stretch the graph vertically by ratio a ($a \neq 0$), you would graph the curve $y = a \cdot f(x)$.

$$a \cdot f(x) = \frac{a}{x} = \frac{1}{\frac{1}{a}x} = f\left(\frac{1}{a}x\right)$$

This shows that a vertical stretch by ratio a is equivalent to a horizontal stretch by the same ratio. Although you could say that the graph is stretched vertically or horizontally by ratio a, it would not be correct to say that it is stretched vertically *and* horizontally by ratio a. That would be a dilation by ratio a, which this is not.

This activity addresses a class of rational functions in this form:

$$g(x) = \frac{cx + d}{ex + f}, \quad \text{where } e \neq 0$$

Understanding the properties of the function and the shape of its graph becomes easier when you rewrite the function:

$$g(x) = \frac{a}{x - h} + k$$

In this form you can create the function by applying three transformations (corresponding to a, h, and k) to the reciprocal function $f(x) = 1/x$.

1. Open **Rational Functions Present.gsp.** The curve on the screen is the graph of a rational function.

2. Guide students through these steps to change the form of the function:

$$\frac{4x - 10}{x - 3} = \frac{4(x - 3) + 2}{x - 3} = \frac{2}{x - 3} + 4$$

3. If $f(x) = 1/x$, then this function is $2f(x - 3) + 4$. You should be able to get the same graph by putting the parent function (the reciprocal curve $y = 1/x$) through three transformations. Stretch it vertically by a ratio of 2, translate it right 3 units, and translate it up 4 units.

4. Press the *Show Parent Function* button. This shows the graph of $y = 1/x$.

5. Show the three transformations by pressing the corresponding buttons (*Stretch, Horizontal Translation,* and *Vertical Translation*) in order, pausing at each step to check for understanding. You will also be able to see the transformations of the asymptotes. Press *Reset* when you finish.

Do not change the order of the transformations.

6. Above and below the equation in the upper-left corner are four parameters corresponding to the parameters in the rational function. Edit the parameters in order to change the equation. Try the equations below. Have the class change the form of the function as you did in the first example. (The solutions are below the problems.) Repeat step 5 to see the transformations.

To edit a parameter, either double-click it or select it and press the plus (+) or minus (–) key.

a. $y = \dfrac{-3x - 12}{x + 5}$

$\quad = \dfrac{3}{x + 5} - 3$

b. $y = \dfrac{4x - 17}{4x - 16}$

$\quad = \dfrac{-0.25}{x - 4} + 1$

c. $y = \dfrac{15x + 103}{5x + 35}$

$\quad = \dfrac{-0.4}{x + 7} + 3$

Modeling Linear Motion: An Ant's Progress

You notice an ant walking slowly across a piece of graph paper on your desk. You start your stopwatch the moment the ant is at point $(-4, -5)$; one minute later, it has reached point $(-1, -3)$. You find yourself wondering when the ant, if it keeps up this pace, will cross the y-axis and, after that, the x-axis.

Spend some time pondering this situation. For now, focus on understanding and picturing the situation rather than finding exact solutions. If you do come up with possible answers, write them down.

INVESTIGATE

Now that you have an understanding of the situation, let's look at how algebra and Sketchpad can help you explore the ant's walk.

Consider the following two equations, which together give the ant's location at any time t:

$$x = -4 + 3t \quad \text{and} \quad y = -5 + 2t,$$

where t is the number of minutes the ant has walked.

Equations like these are called *parametric equations,* and the variable t is called a *parameter.*

Q1 Use the equations above and the given values of t to fill in the table. To do this, substitute the t-values into the equations to find the ant's (x, y) locations at the given times.

For example, for $t = 0$:

$$x = -4 + 3(0) = -4$$
$$y = -5 + 2(0) = -5$$

which confirms that the ant is at point $(-4, -5)$ at time $t = 0$.

t	0	1	2	3	4
x	-4				
y	-5				

Q2 Plot each (x, y) pair from Q1 on a coordinate grid (either in Sketchpad or on your paper) to show where the ant is after 0, 1, 2, 3, and 4 minutes.

Q3 What would the two parametric equations be if the ant started at point $(-2, 0)$ and reached point $(0, 1)$ after one minute?

SKETCH

1. Open **Linear Motion.gsp.**

You'll see a coordinate grid with a red line representing time, the parametric equations, and, of course, an ant.

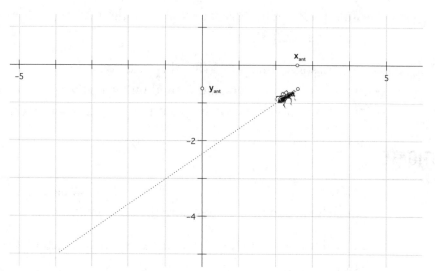

2. Familiarize yourself with the sketch by moving the ant along its path (by dragging marker *t*) and also by pressing the buttons in the sketch. Compare the ant's locations in the sketch with the positions you plotted by hand.

When the ant crosses the *x*-axis, the *y*-value must be zero. Substitute zero in the parametric equation for *y*.

Q4 Use the parametric equations themselves to determine the exact times when the ant crosses the *x*- and *y*-axes. Use the sketch to check your answers.

Q5 What are the *x*- and *y*-intercepts of the line the ant is following?

Press and hold the **Straightedge** tool to choose the **Line** tool from the pop-up menu.

Q6 With the trace showing, draw a line over it as accurately as you can. Select the line and measure its equation. The measurement appears in slope-intercept form. How are the slope and intercept related to the two parametric equations?

To edit the parametric equations, double-click them. To enter *t* into the Calculator dialog box that appears, click on its measurement in the sketch.

Q7 Consider the following motion: The ant starts at $(5, 1)$ and heads toward $(-7, 19)$. Edit the parametric equations to model this situation, and test the model by pressing the action buttons.

Q8 How would the equations in Q6 change if you insisted that the ant travel the same speed as at the beginning of the activity? Edit the sketch to model this situation, test the model by pressing the action buttons, and write your equations on your paper.

Exploring Algebra 2 with The Geometer's Sketchpad
© 2007 Key Curriculum Press

Modeling Linear Motion: An Ant's Progress

Objective: Students model linear motion using parametric equations.

Student Audience: Algebra 2

Prerequisites: None

Sketchpad Level: Easy

Activity Time: 20–30 minutes

Setting: Paired/Individual Activity (use **Linear Motion.gsp**) or Whole-Class Presentation (use **Linear Motion Present.gsp**)

Although this activity covers an advanced topic, parametric equations of lines, it is an accessible and fun activity for students to do. It's especially valuable for the connections between the parametric equations and the slope-intercept form of lines.

INVESTIGATE

Q1

t	0	1	2	3	4
x	-4	-1	2	5	8
y	-5	-3	-1	1	3

Q2 Students should plot the following points: $(-4, -5)$, $(-1, -3)$, $(2, -1)$, $(5, 1)$, and $(8, 3)$

Q3 $x = -2 + 2t$ and $y = 0 + 1t$ (or $y = t$)

SKETCH

Q4 When the ant crosses the y-axis, $x = 0$.

$$-4 + 3t = 0$$

$$t = \frac{4}{3} \text{ (1 minute, 20 seconds)}$$

When it crosses the x-axis, $y = 0$.

$$-5 + 2t = 0$$

$$t = \frac{5}{2} \text{ (2 minute, 30 seconds)}$$

Q5 You can find the intercepts from the answers to Q4. The ant crosses the y-axis when $t = 4/3$.

$$y = -5 + 2\left(\frac{4}{3}\right) = -\frac{7}{3}$$

It crosses the x-axis when $t = 5/2$.

$$x = -4 + 3\left(\frac{5}{2}\right) = \frac{7}{2}$$

Q6 The equation is approximately $y = 0.67x - 2.33$. Written using fractions, this is $y = 2/3x - 7/3$. (Depending on how accurately students have drawn their lines, their results for the slope and y-intercept may differ by a few hundredths.) The slope (0.67, or 2/3) is the ratio of the coefficients of t in the equations for y and x. The y-intercept is the value of y when x is zero (from Q4, at $t = 4/3$). So the y-intercept is

$$y = -5 + 2t = -5 + 2(4/3) = -5 + 8/3 = -7/3$$

This matches the measured value of -2.33.

Q7 Since the speed of the ant is not given, there are many correct answers, but these are the most likely:

$$x = 5 - 12t, \quad y = 1 + 18t$$

The ant must start at $(5, 1)$, and the path must have a slope of $-3/2$. The slope is the ratio of the t coefficients.

Q8 In the first situation, the ant was going right 3 units and up 2 units every minute. If it goes left 2 units and up 3 units per minute, it will be traveling at the same speed, and the slope of its path will satisfy the conditions of this new problem.

$$x = 5 - 2t, \quad y = 1 + 3t$$

Modeling Linear Motion: An Ant's Progress

Parametric equations use x and y as separate functions of a third variable, t. This idea is a considerable leap. Typically, you will introduce t as time. But with this presentation you have an advantage over chalkboard lessons. Here the variables actually do change over time.

1. Open **Linear Motion Present.gsp.** Describe the scenario:

   You have a sheet of graph paper on your table, and you see an ant slowly walking across the grid. Unlike most ants, this one appears to know where it is going. It is walking in a straight line and at a steady pace.

2. Press *Show Ant's Motion* and ask students to observe the ant's journey.

3. Explain that you are going to trace the first minute of the trip. Point out the measurement t, which represents the time in minutes. Press in order *Reset, Show Traces,* and *Advance 1 minute.*

> The actual speed of the animation will be much faster. You will need to emphasize that the time units are minutes.

Q1 What were the coordinates of the ant at time zero? $(-4, -5)$ What are its coordinates at time 1? $(-1, -3)$

Q2 Tell students to think about the x-coordinate only. It started at -4, and after 1 minute, it was -1. What will it be after 2 minutes? (2) What will it be after 3 minutes? (5)

Q3 So the x-coordinate of the ant starts at -4, and advances 3 units every minute. How can students express x in terms of t? $(x = -4 + 3t)$

Q4 Using the same reasoning, what is y in terms of t? $(y = -5 + 2t)$

4. Press *Show Equations* to show the same equations derived above. Explain the concept of parametric equations. Rather than y being a function of x, you have both x and y as functions of parameter t.

Q5 Press *Show Ant's Motion,* and let it run long enough that students can clearly see the linear path. What is the slope of the path? How can they derive it from the parametric equations? (The slope is 2/3, which is the ratio of the coefficients of the t terms in the equations.)

> After getting answers to Q6 and Q7 (right or wrong), double-click the equations to make the changes.

Q6 Ask, "How can you change the equations to make the ant go twice as fast without changing its path?" (Multiply both of the t terms by two. $x = -4 + 6t$, $y = -5 + 4t$)

Q7 A different ant starts at $(-6, 2)$, and after 1 minute it is at $(-1, 0)$. What parametric equations model its motion? $(x = -6 + 5t, y = 2 - 2t)$

7

Trigonometric Functions

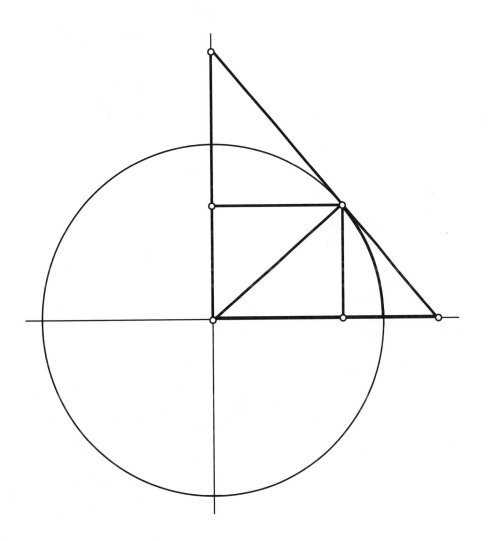

Right Triangle Functions

Within a right triangle there are many important and useful ratios. You may have used *SOH CAH TOA* as a way of remembering the names of certain right triangle ratios:

    SOH: The ratio for *Sine* is *Opposite* over *Hypotenuse*.

    CAH: The ratio for *Cosine* is *Adjacent* over *Hypotenuse*.

    TOA: The ratio for *Tangent* is *Opposite* over *Adjacent*.

In this activity you'll explore these ratios in triangles of different shapes and sizes.

RATIOS

1. Open **Right Triangle Functions.gsp.** Drag each point and observe its effect on the size and shape of the triangle.

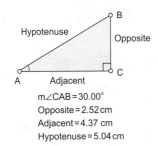

$m\angle CAB = 30.00°$
Opposite = 2.52 cm
Adjacent = 4.37 cm
Hypotenuse = 5.04 cm

2. Use the **Text** tool to label the sides of the triangle *Opposite*, *Adjacent*, and *Hypotenuse* based on their position relative to $\angle CAB$.

3. Measure $\angle CAB$ and the length of each side.

4. Calculate the sine ratio for $\angle CAB$ by dividing the appropriate sides of the triangle. Label the resulting ratio *sin A*.

*To calculate the ratio, choose **Measure | Calculate** and enter each side into the calculation by clicking its measurement in the sketch.*

*Double-click the **Text** tool on the resulting calculation to change its label.*

5. Drag *B* and observe its effect on sin *A*. Use the button to make $m\angle CAB = 30°$.

Q1 What is sin 30°?

Q2 If you make the triangle larger without changing $m\angle CAB$, do you think sin *A* will increase, decrease, or stay the same? Give a reason for your prediction.

Q3 Change the size of the triangle by dragging the *Length of Hypotenuse* slider. What happens to $m\angle CAB$? What happens to sin *A*?

6. Construct the other two ratios based on *SOH CAH TOA*. Label the ratios *cos A* and *tan A*.

What term is used for triangles of different size but the same shape?

Q4 Why do the trigonometric functions stay the same if the triangle is made larger or smaller without changing the angles?

ANGLES UP TO 90°

You have just determined how the triangle's size affects (or does not affect) the values of the ratios. Now you'll make a table and graph to investigate how the angle affects the ratios.

7. Place $m\angle CAB$, $\sin A$, $\cos A$, and $\tan A$ in a table by selecting all four values in order and choosing **Graph | Tabulate.**

8. Change the angle and watch the numbers change. Double-click the table to make the first row permanent. Continue dragging B and double-clicking the table until you have ten rows of values for different angles between 0° and 90°.

9. Press *Show Axes*. Then select the table and choose **Graph | Plot Table Data.** Choose $m\angle CAB$ for x and $\sin A$ for y.

Q5 On your paper, sketch the shape of the plotted points.

To plot a point, select two measurements in order and choose **Measure | Plot As (x, y).**

10. To get a more complete graph of the relationship, plot the point ($m\angle CAB$, $\sin A$). Turn on tracing for the plotted point.

11. Fill in the gaps of your graph by dragging B to change the angle from 0° to 90°.

To trace a point, select it and choose **Display | Trace Plotted Point.**

12. Following the same steps, plot both ($m\angle CAB$, $\cos A$) and ($m\angle CAB$, $\tan A$). Sketch and label all three graphs on your paper.

Q6 Describe the behavior of each graph. Is it increasing, decreasing, or constant? For what angles is it changing quickly? For what angles is it changing slowly?

You can use the buttons to set the angle exactly.

Q7 Does $\cos 90° = \cos 30° + \cos 60°$? Use your cosine graph to explain your answer.

As a hint, set angle A to 60°. What is angle B? Look at the sides involved for $\cos A$ and $\sin B$.

Q8 Why does $\cos 60° = \sin 30°$? Use your answer to find a second pair of angles for which the cosine of one is equal to the sine of the other.

Q9 What happens to the tangent when the angle is 90°?

EXPLORE MORE

Since a right triangle cannot have an obtuse angle, can $\sin 150°$ exist? What would the triangle look like if you could drag B so that $m\angle CAB > 90°$? What would happen to the ratios?

Q10 Why can a right triangle not have an obtuse angle?

Q11 On page 2 point B is free to move in a complete circle. What happens when you drag the angle past 90°? How could you use the result to define $\sin 150°$?

13. Using Sketchpad's Calculator, calculate $\sin 150°$ and $\sin 210°$.

Q12 How do these values compare to $\sin 30°$?

Q13 Find another angle greater than 180° whose sine is the same as $\sin 210°$.

Right Triangle Functions

Objective: Students calculate various ratios for right triangles. By plotting the resulting values, they see the graphs of the trigonometric functions emerge for angles between 0° and 90°. Finally, students are challenged to think about the values of the ratios for angles beyond 90°.

Student Audience: Algebra 2

Prerequisites: None

Sketchpad Level: Intermediate. Students measure lengths and angles, change object labels, tabulate values, and plot points.

Activity Time: 20–30 minutes

Setting: Paired/Individual Activity (**Right Triangle Functions.gsp**) or Whole-Class Presentation (use **Right Triangle Functions Present.gsp**)

RATIOS

Q1 $\sin 30° = \frac{1}{2}$. If students drag point B to make the angle 30°, the value could be slightly different. By using the button, they will get the exact value.

Q2 Answers will vary. The important thing is that students make predictions before trying it.

Q3 As students change the slider, the size of the triangle changes, but the angle and ratio remain constant.

Q4 The angles for the triangle remain constant as the size of the triangle changes. The resized triangle is similar to the original, so there is a scale factor by which each side has been multiplied to generate the new triangle. Therefore, each ratio must remain constant, because its numerator and denominator have been multiplied by the same factor.

ANGLES UP TO 90°

Q5 Students should end up with a sketch of the sine function between 0° and 90°.

12. It's important that students sketch the graphs themselves on paper, rather than printing out the sketch. The physical act of sketching the shapes helps them to remember the characteristics of the graphs.

Q6 The value of the sine increases from 0° to 90°, quickly at first and then more and more slowly. The value of the cosine decreases from 0° to 90°, slowly at first and then more quickly. The value of the tangent increases from 0° to 90°, very quickly as it approaches 90°.

Q7 No, $\cos 90° \neq \cos 30° + \cos 60°$. A reason can be found by looking at the plot for cosine. Using the plot, you can see that the value of $\cos 30° + \cos 60°$ is definitely greater than the value of $\cos 90°$. In general, there are very few functions where $f(x+y) = f(x) + f(y)$.

Q8 The values of $\cos 60°$ and $\sin 30°$ can be represented in one triangle. If $\angle A$ is 60°, $\angle B$ must be 30° since this is a right triangle. This illustrates that cofunctions of complementary angles are equal, because they refer to the same parts of the triangle.

Q9 The tangent value is undefined. The length of the adjacent side is 0, and division by zero is undefined.

EXPLORE MORE

Q10 If the triangle had an obtuse angle and a right angle, the sum of the angles would be more than 180°.

Q11 When you move B past the vertical to try to make an angle of 150°, you end up with a right triangle with an angle of 30°. (This angle is called the *reference angle* for 150°.) You could use this result to define $\sin 150° = \sin 30°$. (Although this result is correct, $\cos 150° \neq \cos 30°$, because the signs are different. Proper definitions of these functions for angles over 90° is best done by using the unit circle.)

Q12 The Calculator results indicate that $\sin 150° = \sin 30°$ and that $\sin 210° = -\sin 30°$. Students may speculate that the difference is that the opposite side goes up for 150° but down for 210°.

Q13 Another angle with the same sine is 330°.

WHOLE-CLASS PRESENTATION

Use the Presenter Notes and **Right Triangle Functions Present.gsp** to present this activity to the whole class.

Use this presentation to review the right triangle definitions of the trigonometric functions and extend them to angles greater than 90°.

1. Open **Right Triangle Functions Present.gsp.** The measure of $\angle CAB$ is on the screen. Drag point B to show its range.

Q1 For $\angle CAB$, identify the opposite side, adjacent side, and hypotenuse.

2. Press *Show Labels* and *Show Lengths.* Review the sine, cosine, and tangent definitions.

Q2 Set $\angle CAB$ to an arbitrary angle. Which of the three ratios is the greatest? Which is the least? (The answer depends on the angle. Students should be able to make a good guess by looking at the sides.) Press *Show Ratios* to reveal the answers. Repeat this part.

3. Pick another arbitrary angle. Have a student write the three ratios on the board.

Q3 What is the complement of this angle? What are the sine, cosine, and tangent of the complement?

4. Press the *Complement* button and discuss the relationships between the ratios for the original angle and the ratios for the complement.

5. Go to page 2. Again, you can change the angle by dragging a point.

6. Press *sin x* to see the point $(x, \sin x)$. Change the angle to create a trace of the sine graph. Do the same for the cosine and the tangent.

Q4 What is the geometric relationship between the sine and cosine graphs, and how is this explained by the relationship $\sin x = \cos(90° - x)$?

Q5 Why is there no upper limit to the tangent function?

Page 3 introduces trigonometric functions for angles greater than 90°. Discuss the fact that although an angle can be greater than 90°, an angle of that size will not fit into a right triangle. In that case, there is a related acute angle that serves as a *reference angle.* On page 3, $\angle CAB$ is the reference angle for $\angle DAB$.

7. Press *Show Angle CAB* and *Show Angle DAB.* Drag point B around to change the angles.

Q6 How are the sine, cosine, and tangent of these angles related? Students will see that the functions of $\angle DAB$ are sometimes negative, but always have the same magnitude as the corresponding functions of $\angle CAB$. Keep dragging point B and guide the class in determining the domain in which each function is negative.

8. On page 4 trace the function graphs again, this time for all angles between 0° and 360°.

Radian Measure

So you think you know angle measurement? An understanding of degrees is a valuable skill, but there are other ways to measure. Other angle units include points, grads, mils, and dekans. To make matters worse, these units may go by different names in different places. Still worse, they may have the same name, but different definitions.

Above all of these angle units, the radian holds a special place. You can use it to measure angles, of course, but radian measure also describes relationships between certain geometric objects.

WHAT IS A RADIAN?

1. Open **Radians.gsp.** Press the *Go* button and watch the circle radius rotate into a tangent position and then roll around the circle.

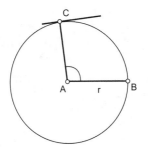

You will use this line segment to measure a central angle of the circle. This will be the basis for defining radian measure for angles.

2. Press *Reset* and then *Home* to stop the animation and return the radius to its tangent position. Measure the radius of the circle and the length of the blue segment.

3. Press the buttons *Show Central Angle*, *Show Arc*, and *1 Radian*. Measure the length of arc *a*.

Q1 The central angle, ∠*BAC*, is now exactly one radian. What do you notice about your three measurements? Explain why they come out this way.

Q2 The measure of the angle, *theta*, is displayed in degrees. Approximately how many degrees are there in one radian?

> Notice that although *theta* is equivalent to *m∠BAC*, it can keep increasing past 360°.

4. Press *Semicircle*. The line segment will continue to roll until it has stepped off half of the circle.

Q3 Using the tick marks to approximate an answer, how many radians are in a semicircle? How many radians will there be in a complete circle?

5. Press the *1 Circle* button to check your last answer.

6. Choose **Edit | Preferences.** Change the Angle Units to **radians.**

Q4 The angle measurement now shows you exactly how many radians are in a circle. But you already knew that, didn't you? Write the formula for the circumference in terms of the radius. Use that along with the definition of a radian to prove that there are exactly 2π radians in a circle.

WHY RADIANS?

So far, you have not seen any good reason for using radians rather than degrees. Actually, we use radians in order to make things easier, not harder.

Do these measurements one at a time. Select one object and choose the appropriate command from the Measure menu.

To change the radius, drag point *B*.

7. Press *Reset* and *Go*. Press *Go* again to stop the animation before the angle makes a complete circle ($0 < theta < 2\pi$).

8. You have a measurement for angle *theta* and a measurement for radius *r*. Use the calculator to find the product *theta · r*.

Q5 What is the arc length in terms of *theta* and *r*? Check your answer with different radii and different angles in the range $0 < theta < 2\pi$. Does your formula always work? Does it work when you use degrees?

The area of a circle sector varies directly with the central angle. You probably are familiar with this formula:

$$\text{sector area} = \frac{\theta}{360°}\pi r^2$$

Q6 Rewrite the above formula using radians instead of degrees. Simplify your answer.

9. Select the arc and choose **Construct | Arc Interior | Arc Sector.** Select the sector and choose **Measure | Area.**

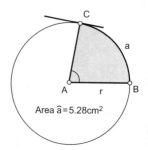

Area $\widehat{a} = 5.28 \text{cm}^2$

Q7 Using your formula from Q6, calculate the area of the sector. Does it match your measurement in all cases?

DISCUSS

Q8 When using radians, Sketchpad automatically expresses angle measurements in multiples of π. This is a common practice. Why?

Q9 It is also common practice (not used by Sketchpad) to write radian angle measurements without writing any units at all. Why is that?

Q10 In spite of the radian advantages you have seen here, degrees are more common in practical applications. What advantages do degrees have?

Radian Measure

Objective: Students explore the relationship between the length of a circular arc, its radius, and its central angle measured in radians.

Student Audience: Geometry/Algebra 2/Precalculus

Prerequisites: Students need no previous experience with radians. However, if this is an introduction, it would be best to show them a sketch of a one-radian sector and explain that the arc length is equal to the radius.

Sketchpad Level: Intermediate. Students must do a few measurements and calculations.

Activity Time: 20–30 minutes

Setting: Paired/Individual Activity (use **Radians.gsp**) or Whole-Class Presentation (use **Radians Present.gsp**)

WHAT IS A RADIAN?

Q1 All three measurements (the radius of the circle, the length of the blue segment, and the length of arc *a*) are the same. This is because the blue segment started out as a radius of the circle, and when it rolled along the circle, it measured out an arc that is the same length as it is.

Q2 1 radian $\approx 57.30°$

Q3 A semicircle has a central angle of exactly π radians. If students are not aware of that fact, they should be able to give an estimate between 3.0 and 3.3. There are 2π radians in a complete circle.

Q4 Start with the circumference formula:

$$\text{circumference} = 2\pi r$$

This means that there are 2π radius lengths in the circumference, so the angular measure is 2π radians.

WHY RADIANS?

Q5 The measurements should verify this fundamental relationship:

$$\text{arc length} = \theta r$$

The formula will work for any angle in the given range, and for any radius. It will not work when the angle units are degrees.

Students may notice some conflicts with the units. The arc length will be in centimeters, but the θr calculation will be in radians $\cdot$ centimeters. Leave that discrepancy for the discussion.

Q6 When degrees are converted to radians, you get this simpler formula:

$$\text{area} = \frac{\theta}{360°}\pi r^2 = \frac{\theta}{2\pi}\pi r^2 = \frac{\theta r^2}{2}$$

Q7 The formula works for any angle between 0 and 2π, and for any radius, but it does not work when the angle units are degrees. Again, the number values agree, but there is an apparent discrepancy with the units.

DISCUSS

The discussion questions will be more helpful if the entire class works together. Here are some suggested points.

Q8 Although radian angle measurement is useful, an angle of one radian has no great significance. The really useful angles ($180°, 90°, 60°$) have measures that are irrational numbers when expressed in radians, so they cannot be expressed exactly with a decimal expansion. However, we can express them as simple fraction multiples of π ($\pi, \pi/2, \pi/3$).

Q9 A radian measurement is an angle measurement, but you can just as well think of it as the ratio of the lengths of an arc and its radius. Since it is a ratio of two linear measurements, it has no units. An advantage of this concept is that it clears up the unit discrepancy that appeared in Q5 and Q7.

Q10 The number of radians in a circle is 2π, an irrational number. It is impossible to divide a circle into an integral number of radians. It is also impossible to do this with a tenth, hundredth, thousandth, or any other fraction of a radian. Instruments that measure angles (protractors, compasses, theodolites, sextants) need to have divisions that are all the same, so the angle unit must divide the circle evenly.

One obvious solution might be simply to graduate the instruments in some fraction of π radians. In fact, that is exactly what we do ($1° = \pi/180$ radians).

1. Open **Radians Present.gsp.** Press the *Go* button.

As the radius segment rolls around the circle, explain that students can think of a radian as the angle that corresponds to one length of the radius being laid out along the circumference of the circle.

Q1 How long is the blue segment? (It's equal to the radius of the circle.)

Q2 What do the red ticks mark off? (Each tick marks a distance of one radius and an angle of one radian.)

2. Press *Reset,* then *1 Radian.* The animation stops after marking off one radian.

Q3 About how many degrees are there in one radian? To coax a good guess, point out the triangle formed by points *A, B,* and *C.* You can think of arc *BC* as a side of the triangle. That would make it an equilateral triangle, but one of the sides is not straight. Would that make $\angle BAC$ greater than 60° or less than 60°?

3. Press *Show Central Angle.* It will show that *theta* $\approx 57.30°$.

4. Press *Semicircle.*

Q4 Count the tick marks. How many radians are there in a semicircle? (a little more than 3)

To change the angle units, choose **Edit | Preferences.**

5. Change the Angle Units to radians. Angle *theta* appears as 1π radians.

6. So a semicircle has π radians. That means that a circle must have 2π radians. Press *1 Circle* to confirm that.

Q5 But you already knew that, didn't you? What is the circumference in terms of *r*? ($2\pi r$) So how many times will the radius go into the circumference? (2π) So how many radians are there in a circle? (2π)

7. Press *Reset,* then press *Go.* Press *Go* again to stop the animation with *theta* somewhere between 0 and 2π. Press *Show Arc.*

When entering numbers that are on the screen, click on the measurement itself.

8. Here is something you can do with radians, but not with degrees. Choose **Measure | Calculate.** Enter *theta* $\cdot r$. Compare the calculation with the measured length of the arc. Try it with several different values of *theta* and *r*.

Q6 The formula for area of a sector is $\frac{\theta}{360°}\pi r^2$. Convert the 360° to radians and simplify. What is the new formula? $\left(\frac{\theta r^2}{2}\right)$

9. Press *Show Sector.* Use the Sketchpad Calculator and enter *theta* $\cdot r^2/2$. Compare the answer to the measured arc length.

Unit Circle Functions

There are several different ways of defining trigonometric functions like sine and cosine. One set of definitions is based on right triangles, but right triangle definitions are limited to angles between 0 and $\pi/2$. (Recall that $\pi/2 = 90°$.) In this activity you'll use a *unit circle* (a circle with a radius of exactly one unit) to define trigonometric functions for any possible angle, even beyond 2π.

CONSTRUCT A UNIT CIRCLE

Start by creating a coordinate system, constructing a unit circle, and making some measurements.

Use the Units panel of Preferences to set units and precision, and use the Color panel to turn on trace fading.

1. In a new sketch, set the Angle Units to **radians,** set the Precision for slopes and ratios to **thousandths,** and turn on trace fading. Use the Preferences dialog box to make all three of these settings, by choosing **Edit | Preferences.**

2. Choose **Graph | Show Grid** and resize the axes (by dragging the number on one of the tick marks) so that the maximum *x*-value is between 6 and 7.

3. Label the origin *A* and the unit point *B* by selecting them in order and choosing **Display | Label Points.**

4. Construct a unit circle. With points *A* and *B* still selected, choose **Construct | Circle By Center+Point.**

5. Construct a point on the circle and label it *C.* (Be sure you don't construct it where the circle intersects one of the axes.)

6. Measure the *x*- and *y*-coordinates of this new point separately. Choose **Measure | Abscissa (x)** and **Measure | Ordinate (y).**

Use the **Line** tool, or else select points *A* and *C* and choose **Construct | Line.**

7. Construct a line through the origin and the point that you just labeled.

8. Measure the slope of this line by choosing **Measure | Slope.**

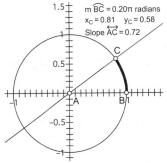

To construct the arc, select the circle and points *B* and *C* in order. Then choose **Construct | Arc On Circle.** Choose **Display | Line Width | Thick** to make it thick.

9. On the circle, construct an arc that begins at the *x*-axis (at unit point *B*) and goes counter-clockwise to point *C.* Make the arc thick.

10. With the arc still selected, measure its arc angle.
    Drag point *C* around the circle and observe how all four measurements behave.

   Q1 What are the largest and the smallest values you observe for each measurement? Where do you find these largest and smallest values?

PLOT YOUR MEASUREMENTS

To explore how the measured quantities depend on the position of point C, you'll plot each measurement using the arc angle as the independent variable.

To plot the point, select in order the independent variable (the arc angle) and the dependent variable (the y-coordinate), and choose **Graph | Plot As (x, y).**

11. Plot the y-coordinate of point C as a function of the arc angle. With the plotted point selected, choose **Display | Trace Plotted Point.**

Q2 Examine the trace that appears as you drag point C around the circle. Describe its shape as you drag point C through the four quadrants. Do you recognize this graph? Which trigonometric function is this?

You may want to turn off tracing for your first plotted point before looking at the second one.

12. Plot the x-coordinate of point C as a function of the arc angle. Turn on tracing for this plotted point, and then drag C to observe how it behaves.

Q3 Describe the shape of this trace as you drag point C through the four quadrants. Which trigonometric function is this?

13. Plot the slope of the line as a function of the arc angle. Turn on tracing, drag C, and observe the result.

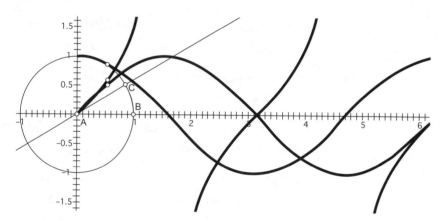

Q4 Describe the shape of this trace as you drag C through the four quadrants. Which trigonometric function is this?

Q5 Calculate the value of y_C/x_C. Compare this value to the value of the slope while you drag point C. What do you notice? Explain your observations.

EXPLORE MORE

Q6 Through point B, create a tangent to the unit circle by constructing a line perpendicular to the x-axis. Construct the intersection of this tangent line with the line through points A and C, and measure the coordinate distance from the point of tangency to this intersection. How does this measurement compare with other measurements you have made? How does this measurement help explain the name of one of the trigonometric functions?

Exploring Algebra 2 with The Geometer's Sketchpad
© 2007 Key Curriculum Press

Unit Circle Functions
continued

The unit circle allows you to use angles greater than $\pi/2$, but by using an arc, you are still limited to positive angles less than 2π. To use angles outside this domain, change point C so it's a rotated image of point B.

14. Create an angle parameter by choosing **Graph | New Parameter.** Name the angle *theta* and set its units to radians.

15. Mark point A as the center of rotation by selecting it and choosing **Transform | Mark Center.**

16. Rotate point B by the value of *theta* by selecting it and choosing **Transform | Rotate.** When the Rotate dialog box appears, click *theta* in the sketch to use the parameter as the angle of rotation.

17. Split point C from the circle and merge it to the rotated image. (Select C and choose **Edit | Split Point From Circle.** Then select both C and the rotated image, and choose **Edit | Merge Points.**)

The value of the parameter changes by $\pi/36$ (the equivalent of 5°) every time you press the **+** or **−** key. By using only these keys to change the parameter, you can keep its value at round numbers.

Q7 Set *theta* to zero, and then press the **+** key repeatedly to change the angle. Record at least four angles for which the slope is zero. Then use the **−** key to find two more angles (less than zero) for which the slope is zero.

Q8 Find three different angles for which the y-value of the point on the circle is 1. At least one of your angles should be negative.

Q9 Find one angle for which the x-value of the point on the circle is -1. Then write down five more such angles, without actually trying them.

Q10 Find one angle for which the y-value of the point on the circle is 1/2. Then write down five more such angles.

Unit Circle Functions

Objective: Students use a unit circle to define the trigonometric functions. They construct a point on the unit circle, measure the coordinates of the point, and graph the coordinates as a function of the arc angle from the positive *x*-axis to the point, producing plots of the sine and cosine functions. They also plot the slope of the line from the origin through the point, producing a plot of the tangent function.

Student Audience: Algebra 2/Precalculus

Prerequisites: None. You can use this activity as the first definition of the trigonometric functions, or you can use it after you've already defined the functions in a triangle.

Sketchpad Level: Intermediate. Students perform the construction and measurements themselves.

Activity Time: 35–45 minutes

Setting: Paired/Individual Activity (no sketch needed) or Whole-Class Presentation (use **Unit Circle Functions Present.gsp**)

Related Activity: Unit Circle and Right Triangle Functions

Most teachers define the trigonometric functions in triangles before introducing the unit circle, but others may prefer to present the unit circle definitions first. This activity supports either way of introducing these definitions.

If you are using the activity as the original definition of the functions, you should introduce it by telling students that the purpose of the activity is to define three new functions. Tell them that the function they will describe in Q2 is the sine function, the function they will describe in Q3 is the cosine function, and the function they will describe in Q4 is the tangent function.

If students have already learned the definitions of these three functions in right triangles and have seen graphs of the functions, they should be able to relate the graphs in Q2, Q3, and Q4 to the functions they already know.

In either case it's important to relate the definitions developed in this activity to the definitions in the right triangle. The activity Unit Circle and Right Triangle Functions is an excellent way to make this connection.

CONSTRUCT A UNIT CIRCLE

10. If students measure $\angle BAC$, the measurement will be shown as a value from $-\pi$ to π. By constructing an arc instead and measuring the arc angle, students end up with a measurement from 0 to 2π, and their graph shows a complete period starting at 0.

Q1 The value of x_C ranges from -1 at the left edge of the circle to $+1$ at the right edge. The value of y_C ranges from -1 at the bottom of the circle to $+1$ at the top. Answers for the slope of line *AC* will vary depending on how close students can drag point *C* to the precise bottom of the circle (where they find the smallest value, probably between -1000 and -100) and the precise top (where they find the largest value, probably between 100 and 1000). The arc angle varies from 0 at the right edge of the circle to 2π, also at the right edge of the circle.

PLOT YOUR MEASUREMENTS

Q2 The trace starts at the origin, curves up to its maximum of 1 as *C* leaves Quadrant I, curves back down to 0 as *C* leaves Quadrant II, continues to -1 as *C* leaves Quadrant III, and finally curves back to 0 as *C* reaches the end of Quadrant IV.

If students have already seen graphs of the trigonometric functions, they should recognize this as the graph of the sine function. If you are using this activity to introduce the functions, you will need to tell them that the *y*-coordinate on the unit circle defines the sine function.

Q3 The trace starts at (0, 1), curves down to 0 as *C* leaves Quadrant I, continues to -1 as *C* leaves Quadrant II, curves back to 0 as *C* leaves Quadrant III, and continues on to 1 as *C* reaches the end of Quadrant IV.

If students have already seen graphs of the trigonometric functions, they should recognize this as the graph of the cosine function. If you are using this activity to introduce the functions, you will need to tell them that the *x*-coordinate on the unit circle defines the cosine function.

Q4 The trace starts at the origin, curves up and shoots off the top of the screen as *C* leaves Quadrant I, reappears from the bottom of the screen and curves up to 0 as *C* leaves Quadrant II, continues once more off the top of the screen as *C* leaves Quadrant III, and reappears again at the bottom of the screen and curves back to 0 as *C* reaches the end of Quadrant IV.

If students have already seen graphs of the trigonometric functions, they should recognize this as the graph of the tangent function. If you are using this activity to introduce the functions, you will need to tell them that the slope from the origin to the point on the unit circle defines the tangent function.

Q5 The calculation y_C/x_C gives the same result as the slope measurement, because it corresponds to the definition of slope: rise/run or $\Delta y/\Delta x$. Use students' answers to this question to make the point that the tangent function can be defined in three equivalent ways: as the slope, as y/x, or as $\sin\theta/\cos\theta$.

EXPLORE MORE

Q6 The coordinate distance from the point of tangency to the intersection is the same as the slope measurement. Applying the slope formula to the point of intersection (x, y) and the origin $(0, 0)$ results in y/x. Because the x-coordinate along the tangent line is 1, the y-coordinate of the intersection must be equal to the slope; that is, it's equal to the value of the tangent function. Thus, the length of the tangent from the point of tangency $(1, 0)$ to this intersection is the value of the tangent function. It's this fact that originally gave the tangent function its name.

The unit circle extends the domain of the trigonometric functions to 2π; but to go beyond 2π, students need to consider the possibility of wrapping around the unit circle more than once. The remaining three questions require them to do so, and also to consider negative angles as angles measured in the opposite (clockwise) direction.

Q7 Angles for which the slope (tangent function) is 0 include $0, \pi, 2\pi$, and 3π. The angle $-\pi$ is the first angle less than 0 for which the slope is 0. The general form is $n \cdot \pi$ for any integer n.

Q8 The y-value (sine function) of the point on the circle is 1 for positive angles $\pi/2, 5\pi/2, 9\pi/2$, and so forth. It's also 1 for negative angles $-3\pi/2, -7\pi/2$, and so forth. The general form is $\pi/2 + 2\pi n$ for any integer n.

Q9 The x-value (cosine function) of the point on the circle is -1 for positive angles $\pi, 3\pi, 5\pi$, and so forth. It's also -1 for negative angles $-\pi, -3\pi, -5\pi$, and so forth. The general form is $\pi + 2\pi n$ for any integer n.

Q10 The y-value (sine function) of the point on the circle is 1/2 for positive angles $\pi/6, 5\pi/6, 13\pi/6, 17\pi/6$, and so forth. The general forms are $\pi/6 + 2\pi n$ and $5\pi/6 + 2\pi n$ for any integer n.

WHOLE-CLASS PRESENTATION

Use the Presenter Notes and **Unit Circle Functions Present.gsp** to present this activity to your class.

Use this presentation to define the trigonometric functions based on the unit circle. You can use this as the fundamental introduction to these functions, or you can use it after presenting the definitions in right triangles.

PRESENT

1. Open **Unit Circle Functions Present.gsp.**

Q1 Ask students for the radius of the circle. (1)

2. Press the *Animate C* button to move *C* around the circle. Make sure that students notice arc *BC.*

Q2 Show the arc angle measurement, and ask students to determine the smallest and largest values for this measurement. (0 and 2π)

Q3 Stop the animation and ask students what the arc length is. (It's the same as the arc angle, because the radius is 1.)

Do not actually measure the arc length. The measured length would not be based on the coordinate system unit.

Q4 Restart the animation and show the *y*-coordinate. Ask students to observe and determine the smallest and largest values, and where they occur. (The value of y_C ranges from -1 at the bottom of the circle to $+1$ at the top.)

Q5 Tell students you are about to plot the *y*-coordinate as a dependent variable. Ask them what measurement to use as the independent variable. (arc angle *BC*)

3. Stop the animation and press *Show Plotted y-value* to see the plotted point. Drag *C* so students can verify that the plotted point matches their observations from Q4. Then restart the animation to get a smooth plot of the traced point.

Q6 What trigonometric function does this plot represent? (It is the sine function.)

4. Repeat the same actions and questions for the *x*-value (resulting in a plot of the cosine function) and the slope measurement (resulting in a plot of the tangent function).

5. Ask students for the domain of the plots they have seen. (0 to 2π) Then go to page 2 and use the buttons to show how the domain can be extended to negative angles and to angles greater than 2π.

Unit Circle and Right Triangle Functions

There are several different ways to define trigonometric functions like sine and cosine. One set of definitions is based on right triangles, and another set is based on a *unit circle* (a circle with a radius of exactly one unit). In this activity you'll explore the relationship between these two ways of defining trigonometric functions.

THE UNIT CIRCLE

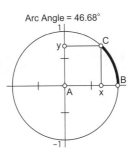

Arc Angle = 46.68°

1. Open **Unit Circle Right Triangle.gsp.** Measure the arc angle of arc *BC* on the unit circle. Label the measurement *Arc Angle.*

Q1 Drag point *C* around the circle and observe the angle measurement. What are the smallest and largest values that you observe? Leave *C* in Quadrant I when you finish.

2. Measure the *y*-coordinate of point *C* and label it *sin in circle.* Measure the *x*-coordinate and label it *cos in circle.*

3. Construct a line through *A* and *C*, and measure the slope of the line. Label this measurement *tan in circle.*

Q2 Drag *C* again. What are the smallest and largest values that you observe for the sine, cosine, and tangent of the arc angle? At what angles do these values occur?

THE REFERENCE TRIANGLE

Ratios of triangle sides provide another way to define trigonometric functions. You can use the mnemonic *SOH CAH TOA* to recall the ratios:

  SOH: The ratio for *Sine* is *Opposite* over *Hypotenuse.*
  CAH: The ratio for *Cosine* is *Adjacent* over *Hypotenuse.*
  TOA: The ratio for *Tangent* is *Opposite* over *Adjacent.*

To measure ∠*E*, select points *D*, *E*, and *F.* Then choose **Measure | Angle.**

4. Measure ∠*E* for the right triangle.

5. Measure the *Adjacent* side by selecting points *D* and *E* and choosing **Measure | Coordinate Distance.** Do the same for the other two sides. Label your measurements *Opposite, Adjacent,* and *Hypotenuse.*

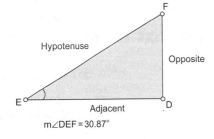

m∠DEF = 30.87°

6. Use Sketchpad's Calculator to calculate *Opposite/Hypotenuse.*

Q3 According to the *SOH CAH TOA* mnemonic, to which trigonometric function does this calculation correspond?

7. Label your calculation *sin in triangle*. Calculate each of the other two ratios and label them appropriately.

Q4 Drag point *F*. What are the smallest and largest values that you observe for the sine, cosine, and tangent in the right triangle? At what angle do the maximum and the minimum occur for each?

COMPARE THE DEFINITIONS

To compare these definitions, you'll combine the two models.

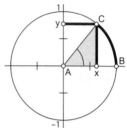

8. Select points *A* and *E*, and choose **Edit | Merge Points**. Also merge points *C* and *F*. The right triangle is now attached to the inside of the unit circle.

Q5 Drag point *C* and observe the two angle measurements (the arc angle and the angle in the triangle). When do these measurements agree? When do they disagree?

Q6 Drag point *C* and observe the two sine measurements. Explain why the values are equal in certain quadrants but not in others.

Q7 When do the two cosine measurements agree, and when do they disagree? Why?

Q8 When do the two tangent measurements agree? Explain.

Q9 Why is the sine of 150° the same value as the sine of 30°? Why is the sine of 210° the opposite of the sine of 30°? (*Hint:* Think about how each relates to either a coordinate or a ratio, and compare these.)

Q10 Describe possible advantages and disadvantages for each method of defining the trigonometric functions.

EXPLORE MORE

Q11 Based on the different definitions, which might be better to determine the flight path of an airplane? The position of a person on a Ferris wheel? The height of a building? Explain.

Q12 Could you always use a single definition? Explain.

Q13 Drag point *C*. What happens to the tangent at 90°? Explain. What does this mean in terms of the graph of the tangent function at 90°?

Unit Circle and Right Triangle Functions

Objective: Students compare the unit circle definitions of trigonometric functions with the right triangle definitions. They combine the two models and examine similarities and differences that emerge.

Student Audience: Algebra 2/Precalculus

Prerequisites: It's best (though not necessary) if students have seen both sets of definitions of the trig functions. (The Related Activities listed below cover this ground.)

Sketchpad Level: Intermediate. Students measure values, and merge points to combine the two models.

Activity Time: 20–30 minutes

Setting: Paired/Individual Activity (use **Unit Circle Right Triangle.gsp**) or Whole-Class Presentation (use **Unit Circle Right Triangle Present.gsp**)

Related Activities: The definitions are introduced in Right Triangle Functions and in Unit Circle Functions.

THE UNIT CIRCLE

Q1 The value of *Arc Angle* ranges from 0° to 360°.

Q2 The sine ranges from −1 to 1, at angles of 270° and 90°, respectively. The cosine ranges from −1 to 1, at angles of 180° and 0°, respectively. The tangent has no limit in either direction, but the measurement is limited by the resolution of the objects on the screen.

THE REFERENCE TRIANGLE

Q3 The calculation corresponds to the sine function.

Q4 The smallest value for sine is 0 and occurs at 0°. The largest value is 1 and occurs at 90°. The smallest value for cosine is 0 and occurs at 90°. The largest value is 1 and occurs at 0°. The smallest value for the tangent occurs at 0°. The tangent has no upper limit, and it gets very large as the angle approaches 90°. At both 0° and 90°, the triangle is degenerate, with various points and sides coinciding.

COMPARE THE DEFINITIONS

Q5 The measurements agree only in the first quadrant. In the other quadrants the arc angle is more than 90°, but the angle in the triangle remains between 0° and 90°.

Q6 The definitions agree in Quadrants I and II because the *y*-value is positive there. The definitions disagree in the other two quadrants because the measured length of a line segment is always positive.

Q7 The cosine values agree in Quadrants I and IV, but disagree in Quadrants II and III. In these two quadrants the *x*-value is negative, but the distance measured in the triangle remains positive.

Q8 The tangent values agree in Quadrants I and III. In Quadrant I the coordinates (for the unit circle definition) and the distance measurements (for the right triangle definition) are all positive, so the two functions agree. In Quadrant III both coordinates are negative, so their ratio is positive, matching the right triangle definition. In the other two quadrants one coordinate or the other is negative, resulting in values that the right triangle cannot produce.

Q9 Explanations will vary. This is a good place to introduce the idea of the *reference triangle* within the unit circle and to observe that the opposite side for both 30° and 150° corresponds to the same *y*-value. For 210°, the opposite side corresponds to a negative *y*-value, so the value of sin 210° is the opposite of that of sin 30°.

Q10 Answers will vary. A big advantage of the unit circle method is the ability to work with angles that are beyond 90°. An advantage of the right triangle method is that it's easier to apply when the angle is not in standard position. (Though students don't know this yet, the unit circle method will allow them to explore topics, such as uniform circular motion, which would not be possible with only a right triangle definition.)

EXPLORE MORE

Q11 Answers will vary. Analyzing the flight path of a plane or the position of a person on a Ferris wheel both benefit from using angles beyond 90°. For the height of a building, a right triangle definition is sufficient.

Q12 You could condense the two methods into one by considering the right triangle method to be a special case of the unit circle in Quadrant I.

Q13 At 90° the line *AC* is vertical, so its slope (and the tangent of 90°) is undefined. The result is that the tangent graph has an asymptote at 90°.

Unit Circle and Right Triangle Functions

Presenter Notes

Use this presentation to relate the two ways of defining the trig functions.

THE UNIT CIRCLE

Angles in this sketch are in degrees rather than radians.

Leave point *C* in motion while students answer these three questions.

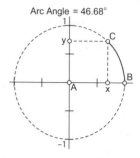

1. Open **Unit Circle Right Triangle Present.gsp.** Show the unit circle and animate point *C*. Show the arc angle.

Q1 Ask students what are the largest and smallest values they observe for the arc angle. (0° to 360°)

Q2 Ask what measurements are needed in the unit circle to define the sin, cos, and tan functions. (in order: *y*, *x*, and equivalently either *y/x* or the slope of *AC*)

Q3 Show these measurements and ask students to observe the largest and smallest values for each of the measurements. Review again which is sin, cos, and tan.

THE REFERENCE TRIANGLE

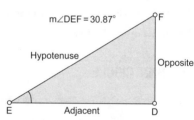

2. Show the right triangle and measure ∠*DEF*.

Q4 Drag point *F* and ask students to observe the largest and smallest values for the angle.

Q5 Show the length measurements and ask students what ratios must be calculated to find the sine, cosine, and tangent.

Q6 Show the ratios and have students confirm which is sine, cosine, and tangent. Drag point *F* and have students observe the largest and smallest values for each ratio.

COMPARE THE DEFINITIONS

3. To compare the definitions, combine the models. Press *Merge Triangle to Circle.*

Q7 Drag point *C* (keeping it in Quadrant I), and ask students to compare the four measurements from each triangle.

Q8 Ask students to make conjectures about what will happen if *C* leaves Quadrant I.

Q9 Drag point *C* slowly through the other three quadrants, and ask students to describe what they observe about each of the four measurements. Encourage them to explain their observations.

Q10 Why does the sine of 150° in the circle have the same value as the sine of 30° in the triangle? Why is sin 210° the opposite of sin 30°?

Q11 Ask for advantages and disadvantages of each way of defining the functions.

Exploring Algebra 2 with The Geometer's Sketchpad
© 2007 Key Curriculum Press

Trigonometric Identities

When you first learned trigonometry, you probably used only acute angles. That's because you can put an acute angle into a right triangle and all of the trigonometric functions will be ratios of sides. Even as you get into more advanced concepts, it can be helpful to keep going back to the right triangle geometry.

ACUTE ANGLES

1. Open the first page of **Trigonometric Identities.gsp.**

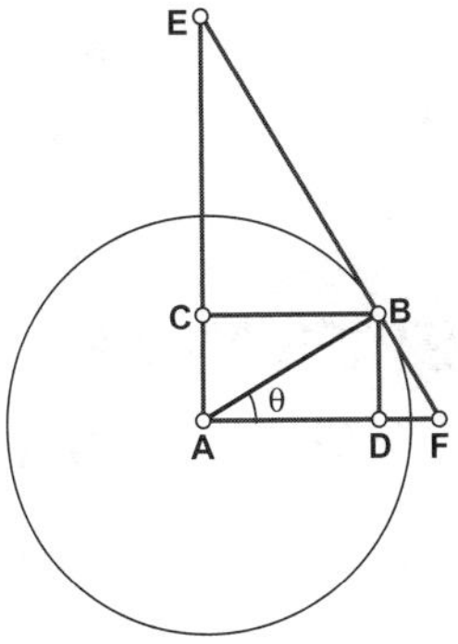

This sketch shows acute angle θ at the center of a unit circle. Controls at the bottom allow you to change the angle or the radius. (When you change the radius, all you change is the scale. The radius of the circle is always one.)

2. To get a feel for the construction, drag the angle control at the bottom of the screen.

Segment AB is the terminal side of angle θ. Segment EF is tangent to the unit circle at point B. The other line segments are either horizontal or vertical.

Q1 Notice that angle θ appears in several places, and there are also a number of right angles. Triangle ABD is a right triangle with angle θ. How many triangles are similar to $\triangle ABD$? Identify as many of them as you can.

Q2 Along the edge of the screen are length measurements for eight line segments. By forming a ratio with AB, you can show that each length is equal to a trigonometric function of θ. Follow this example:

$$\sin \theta = \frac{BD}{AB} = \frac{BD}{1}, \text{ so } \sin \theta = BD$$

For each remaining length measurement, find the corresponding trigonometric function. Some functions appear more than once.

Q3 Lengths of corresponding sides of similar triangles are proportional, so you can write many proportions in this figure. Here is one:

$$\frac{BD}{AB} = \frac{BF}{AF}$$

Using your answers from Q1 and Q2, substitute trigonometric functions into this proportion. What identity do you get?

The measure of angle θ is on the screen, so you can confirm these answers by calculating the trigonometric functions of θ.

Q4 Complete each of the following proportions and use it to derive a trigonometric identity:

a. $\dfrac{AD}{AB} =$

b. $\dfrac{BF}{AB} =$

3. Press *Show Relationship 1*. The squares in this figure show a familiar Pythagorean relationship.

Q5 Using lengths of line segments, express relationship 1 as an algebraic equation. Substitute the trigonometric functions into the equation and state the identity. Do the same for relationships 2 and 3.

OTHER ANGLES

It is a little more difficult to make obtuse angles and reflex angles fit in with the right triangle model. The trick is that you have to use directed distances. Normally the length of a line segment must be positive, but this time you will change the rules to suit your needs. Some of the line segments will represent negative numbers.

4. Press *Any Angle*. Angle θ is no longer restricted to acute angles.

5. Drag the angle control again. Go all the way around the circle.

Q6 When a line segment changes color, check the corresponding measurement. What is the significance of the color changes?

Q7 Now revisit this proportion from Q3:

$$\frac{BD}{AB} = \frac{BF}{AF}$$

Some of these lengths may now be negative, depending on the angle. Will the proportion still be true for all angles? Check at least one angle in each quadrant.

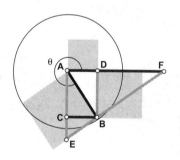

Q8 Use the action buttons to look at the Pythagorean relationships again. Will any of these relationships change when the signs change? Explain.

© 2007 Key Curriculum Press

Trigonometric Identities

Objective: Students review the connections between trigonometric functions and right triangle geometry. They use geometric relationships to justify trigonometric identities.

Student Audience: Algebra 2/Precalculus

Prerequisites: Students should first have a strong understanding of trigonometric function definitions, including the reciprocal functions cosecant, secant, and cotangent.

The activity addresses functions of obtuse angles and reflex angles. If students have only learned functions of acute angles, they should still be able to handle the first section of the activity.

Sketchpad Level: Easy. There are no constructions involved in this activity.

Activity Time: 30–40 minutes

Setting: Paired/Individual Activity (use **Trigonometric Identities.gsp**) or Whole-Class Presentation (use **Trigonometric Identities Present.gsp**).

ACUTE ANGLES

Q1 There are six triangles similar to $\triangle ABD$. They are $\triangle AFB$, $\triangle EAB$, $\triangle BAC$, $\triangle EFA$, $\triangle EBC$, and $\triangle BFD$. Remind students that it is important to have the vertex labels in the correct order.

Q2 You should check in with students here. If they fail to get this part right, they will have little hope of gaining anything from the rest of the activity.

$BD = \sin\theta$ $\qquad\qquad$ $AE = \csc\theta$

$AD = \cos\theta$ $\qquad\qquad$ $BE = \cot\theta$

$AF = \sec\theta$ $\qquad\qquad$ $BF = \tan\theta$

$AC = \sin\theta$ $\qquad\qquad$ $BC = \cos\theta$

Q3 Substitute and simplify:

$$\frac{BD}{AB} = \frac{BF}{AF} \Rightarrow \frac{\sin\theta}{1} = \frac{\tan\theta}{\sec\theta} \Rightarrow \sin\theta = \frac{\tan\theta}{\sec\theta}$$

Q4 There are at least two possible answers for each:

a. $\dfrac{AD}{AB} = \dfrac{AB}{AF} \Rightarrow \dfrac{\cos\theta}{1} = \dfrac{1}{\sec\theta} \Rightarrow \cos\theta = \dfrac{1}{\sec\theta}$

$\dfrac{AD}{AB} = \dfrac{BE}{AE} \Rightarrow \dfrac{\cos\theta}{1} = \dfrac{\cot\theta}{\csc\theta} \Rightarrow \cos\theta = \dfrac{\cot\theta}{\csc\theta}$

b. $\dfrac{BF}{AB} = \dfrac{BD}{AD} \Rightarrow \dfrac{\tan\theta}{1} = \dfrac{\sin\theta}{\cos\theta} \Rightarrow \tan\theta = \dfrac{\sin\theta}{\cos\theta}$

$\dfrac{BF}{AB} = \dfrac{AB}{BE} \Rightarrow \dfrac{\tan\theta}{1} = \dfrac{1}{\cot\theta} \Rightarrow \tan\theta = \dfrac{1}{\cot\theta}$

Q5 Relationship 1:

$$BD^2 + AD^2 = AB^2 \Rightarrow \sin^2\theta + \cos^2\theta = 1$$

Relationship 2:

$$AB^2 + BE^2 = AE^2 \Rightarrow 1 + \cot^2\theta = \csc^2\theta$$

Relationship 3:

$$AB^2 + BF^2 = AF^2 \Rightarrow 1 + \tan^2\theta = \sec^2\theta$$

OTHER ANGLES

Q6 The black line segment, AB, always represents the value 1. Blue line segments represent positive values and red segments represent negative values. You have to use both positive and negative numbers because of the association between the line segments and trigonometric functions.

Students might also notice the geometric pattern in the colors. Horizontal segments are positive (blue) to the right and negative (red) to the left. Vertical segments are positive upward and negative downward. The sign of lengths BE and BF is the opposite of the sign of their common slope.

Q7 When you use directed distances, this proportion remains true for any angle for which the line segments are defined. The same is true for all the proportions from Q4.

Q8 All the Pythagorean relationships hold up. In the equations, every term is squared, so the signs of the substituted values do not matter.

Trigonometric Identities

1. Open **Trigonometric Identities Present.gsp.** Drag the angle control at the bottom so students can see that it is restricted to acute angles.

The circle has a radius of 1, and you can change the scale using the control below. The vertex of the angle is at the center of a unit circle. Make sure the class understands that all of the linear measurements are based on this unit definition. Segment AB is the terminal side of angle θ. Segment EF is tangent to the circle at point B. All the other line segments are either horizontal or vertical.

Q1 Are there any triangles similar to $\triangle ABD$? ($\triangle AFB$, $\triangle EAB$, $\triangle BAC$, $\triangle EFA$, $\triangle EBC$, and $\triangle BFD$)

Q2 Consider BD/AB. That's the sine of θ, but $AB = 1$. What does that make BD? ($BD = \sin\theta$)

The measure of angle θ is on the screen, so you can confirm these answers by calculating the trigonometric functions of θ.

Q3 In fact, each of these remaining measurements corresponds to one of the trigonometric functions. Which is which? Work this out together and keep track of it on the board or in the sketch:

$BD = \sin\theta$	$AE = \csc\theta$
$AD = \cos\theta$	$BE = \cot\theta$
$AF = \sec\theta$	$BF = \tan\theta$
$AC = \sin\theta$	$BC = \cos\theta$

Q4 The similar triangles give us a lot of proportions. Here's one: $BF/AB = BD/AD$. Substitute the trigonometric functions from Q3 for those line segment lengths. What do you get? ($\tan\theta = \sin\theta/\cos\theta$)

Q5 Interpret several other proportions in the same way.

Q6 Press *Show Relationship 1.* By the Pythagorean theorem, $BD^2 + AD^2 = AB^2$. What trigonometric identity follows from that? ($\sin^2\theta + \cos^2\theta = 1$) Do the same with the other Show buttons:

Relationship 2: $AB^2 + BE^2 = AE^2$, so $1 + \cot^2\theta = \csc^2\theta$

Relationship 3: $AB^2 + BF^2 = AF^2$, so $1 + \tan^2\theta = \sec^2\theta$

2. Press *Any Angle.* Now angle θ is no longer restricted to acute angles. Drag the angle control to show this. The measurements are actually directed distances. Their signs change with the corresponding functions, and the line segment colors change to emphasize this.

3. Examine the same identities again. Have the class discuss the question of whether the identities continue to hold for obtuse angles and reflex angles.

Law of Sines

You have already used the sine, cosine, and tangent ratios to find missing parts of triangles. However, the definitions of these functions (involving the ratios of the opposite, adjacent, and hypotenuse) apply only to right triangles. In this activity you'll explore a different set of ratios that you can use in oblique triangles.

MORE RATIOS

1. In a new sketch, construct a triangle. Measure each angle and each side.

To show or change a label, select the object and then choose **Display | Label.**

2. The vertices are automatically labeled *A*, *B*, and *C*. Label the sides *a*, *b*, and *c* according to the vertex that is opposite each side.

Q1 Drag the vertices to make ∠*A* larger than ∠*B*. Which side is longer, *a* or *b*? Is this always true? Write down the measurements from three different examples.

Q2 Drag the vertices to make side *c* longer than *b*. Which angle is larger, ∠*B* or ∠*C*?

3. Calculate the sine of each angle.

To calculate the sine, choose **Measure | Calculate.** Select **sin** from the Function pop-up menu, and then click the angle measurement in the sketch.

4. Calculate the ratio of the length of each side to the sine of the opposite angle.

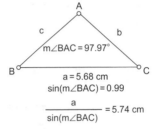

5. Select all three ratios and place them in a table by choosing **Graph | Tabulate.** With the table still selected, choose **Graph | Add Table Data** and choose to add ten entries as the values change.

Q3 Drag the vertices to change the angles and side lengths. What do you observe about the ratios?

Q4 Write your observation as an equation.

Q5 Calculate the reciprocal of each ratio. What do you observe? Write an equation.

These equations are both ways of writing the Law of Sines.

EXPLORE MORE

Q6 Open **Law of Sines Proof.gsp.** Use the labels in the blue triangle to write a formula for sin *A*. Use the labels in the pink triangle to write a formula for sin *B*.

Q7 The length of segment *h* appears in both formulas. Solve both formulas for *h*, and set the results equal to each other.

Q8 What must you do to this equation to complete a proof of the Law of Sines?

Q9 By dragging point *C*, you can move segment *h* so it's outside △*ABC*. Is your proof of the Law of Sines still correct, or must you modify it? Explain.

Law of Sines

Objective: Students construct a triangle, measure some ratios, and find that certain ratios remain equal no matter the shape of the triangle: the Law of Sines. They then use a prepared sketch and apply the triangle definition of the sine function to develop a proof of the Law of Sines.

Student Audience: Algebra 2/Precalculus

Prerequisites: Students must be familiar with the Pythagorean theorem and with the definition of the sine function in a right triangle.

Sketchpad Level: Intermediate. Students must use the Calculator and measure angles and segment lengths.

Activity Time: 15–25 minutes

Setting: Paired/Individual Activity (start with a blank sketch; use **Law of Sines Proof.gsp** for the Explore More section) or Whole-Class Presentation (use **Law of Sines Present.gsp**)

The activity itself does not address the ambiguous case, nor does it provide practice problems for applying the Law of Sines. The Whole-Class Presentation provides both. For this reason you should strongly consider using the presentation sketch as a follow-up to the activity.

MORE RATIOS

Q1 When $\angle A$ is larger than $\angle B$, side a is always longer than side b. Students should record three different sets of all four values ($\angle A$, $\angle B$, a, and b) that they tried.

Q2 When side c is longer than b, $\angle C$ is larger than $\angle B$.

Q3 The ratios remain equal to each other despite changes in the triangle.

Q4 In equation form this is the Law of Sines:

$$\frac{a}{\sin A} = \frac{b}{\sin B} = \frac{c}{\sin C}$$

Q5 The reciprocals of equal nonzero values must also be equal. The second form of the Law of Sines is

$$\frac{\sin A}{a} = \frac{\sin B}{b} = \frac{\sin C}{c}$$

EXPLORE MORE

Q6 By construction, the pink and the blue triangles are always right triangles.

$$\sin A = \frac{h}{b} \qquad \sin B = \frac{h}{a}$$

Q7 Solving for h, the two equations are $h = b \cdot \sin A$ and $h = a \cdot \sin B$. Setting the right-hand sides equal to each other gives $b \cdot \sin A = a \cdot \sin B$.

Q8 Dividing this result by ab gives

$$\frac{\sin A}{a} = \frac{\sin B}{b}$$

Alternatively, dividing the result by $\sin A \cdot \sin B$ gives

$$\frac{a}{\sin A} = \frac{b}{\sin B}$$

Both results are correct statements of the Law of Sines. Once students have written the Law of Sines as the equality of two ratios, ask them to justify adding the third ratio (involving c and $\sin C$) to the equation.

Q9 Once you drag C so h is outside the triangle, one of the two right triangles involves an exterior angle of the triangle. For instance, if you drag C past B, the right triangle DBC no longer contains $\angle ABC$, but now contains the exterior angle at B, which is supplementary to $\angle ABC$. Because $\sin B = \sin(180° - B)$, the Law of Sines is still correct.

WHOLE-CLASS PRESENTATION

Use **Law of Sines Present.gsp** to present this activity to the entire class. Even when students have done the activity on their own computers, use page 4 to demonstrate the ambiguous case.

Page 1 (Intro) describes the contents of the sketch.

Page 2 (Ratios) parallels the main part of the student activity. Use the buttons and directions in the sketch to present this part.

Page 3 (Proof) parallels the Explore More section of the activity.

Page 4 (Ambiguous) explores the ambiguous SSA case.

Pages 5 and 6 (SSA and AAS) make it easy to generate practice problems and their solutions.

Law of Cosines

You know how to use sine, cosine, and tangent functions to solve right triangle problems, and you know how to use the Law of Sines to solve some problems in triangles without right angles. These two methods are very important, but they are not enough to solve all triangle problems. In this activity you'll find another law that allows you to solve problems for which neither of the other methods works.

TRY EXISTING METHODS

Q1 Why can't you use the sine, cosine, or tangent functions directly to find the missing measurements in this triangle?

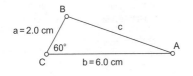

Q2 Write down the Law of Sines and substitute into it the measurements from this triangle. Can you use this equation to solve for the missing side or angles? Explain.

Q3 Can you use the Pythagorean theorem to find the missing side? Give a reason.

The figure at right illustrates the Pythagorean theorem. If you square each side, then $c^2 = a^2 + b^2$. Unfortunately, you can use this theorem only if one of the angles is a right angle. Even so, the Pythagorean theorem is a useful starting place.

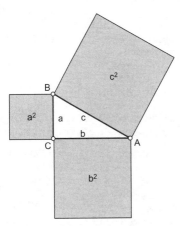

EXTEND THE PYTHAGOREAN THEOREM

1. Open **Law of Cosines.gsp.** Measure the lengths of all three sides of the triangle. Also measure $\angle C$.

2. Calculate the value of a^2 (the area of the square on a). Also calculate b^2 and c^2.

Choose **Measure | Calculate.** To enter a measurement into your calculation, click it in the sketch.

Q4 Calculate $a^2 + b^2 - c^2$. What result do you get? Why?

3. Drag points A and B.

Q5 Why do two of the lengths remain constant? What length does change?

4. You don't need the circles any more, so hide them.

Q6 For what values of $\angle C$ is $a^2 + b^2 - c^2$ positive? For what values is it negative?

GRAPH

It will be useful to figure out exactly how the value of $a^2 + b^2 - c^2$ depends on $\angle C$. To investigate, you'll graph $a^2 + b^2 - c^2$ as a function of $\angle C$.

5. Press *Show Axes*. To plot the current values of the quantities you're investigating, select $m\angle C$ and $a^2 + b^2 - c^2$ in order, and choose **Graph | Plot As (x, y)**.

6. Turn on tracing for the plotted point. Then vary $\angle C$ by dragging A and B.

Q7 Have you graphed a function with a similar shape before? What function?

To trace the point, select it and choose **Display | Trace Plotted Point.**

7. Use $f(x) = \cos(x)$ as the parent function. Plot this function.

Choose **Graph | Plot New Function.** Choose **cos** from the Function pop-up menu. Then click *x* on the keypad.

Q8 How is the graph of $f(x) = \cos(x)$ similar to the trace, and how is it different?

Q9 What transformation could you use to make the function match the trace?

8. Create parameter d and set its value to 1.

To create a parameter, choose **Graph | New Parameter** and use the dialog box that appears to set the name and value of the parameter.

9. Edit function $f(x)$ by double-clicking it. Change it to $f(x) = d \cdot \cos(x)$.

Q10 Change the value of d by selecting it and pressing the + key or − key repeatedly. How does d affect the graph?

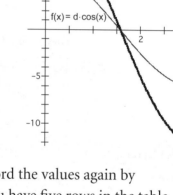

$f(x) = d \cdot \cos(x)$

When you studied stretches and shrinks, you probably used *a* and *b* as parameters. They are called *d* and *e* here to avoid confusion with the sides of the triangle.

Q11 Adjust d until the graph matches the trace. What value of d did you use to do this?

10. Record the values of a, b, and d needed to match the two functions by selecting all three values and choosing **Graph | Tabulate.** Double-click the table to make the first row of values permanent.

If you record a row of values incorrectly, you can remove it by selecting the table and choosing **Graph | Remove Table Data.**

11. Use the sliders to change lengths a and b. Then erase the traces and make a new trace. Adjust d to match the function to the new trace, and then record the values again by double-clicking the table. Repeat this step until you have five rows in the table.

Q12 Examine your table. How can you express d in terms of a and b? Substitute this for d in the function definition. What is the new definition?

Q13 The function is equal to $a^2 + b^2 - c^2$. Solve the resulting equation for c^2. This is the Law of Cosines.

EXPLORE MORE

Q14 What happens to the Law of Cosines if $\angle C$ is a right angle? Substitute the appropriate value and simplify the result. What do you end up with?

Q15 You can also use the Law of Cosines to find an angle when you have all three sides. In that case, how would you write the equation?

© 2007 Key Curriculum Press

Law of Cosines

Objective: Students develop the Law of Cosines by exploring numerically and graphically how the Pythagorean theorem fails for triangles without a right angle. By transforming the cosine function, they find an expression for the discrepancy between c^2 and $a^2 + b^2$. This expression is the term needed to complete the Law of Cosines.

Student Audience: Algebra 2/Precalculus

Prerequisites: Students must be familiar with the Pythagorean theorem, with the graphs of the trigonometric functions, and with stretching and shrinking functions.

Sketchpad Level: Intermediate. Students perform a number of measurements and calculations, plot a point and a function, and tabulate data.

Activity Time: 30−40 minutes

Setting: Paired/Individual Activity (use **Law of Cosines. gsp**) or Whole-Class Presentation (use **Law of Cosines Present.gsp**)

Begin by telling students that they will work out a new way to find missing information in triangles by using the Pythagorean theorem, the graphs of the trigonometric functions, and function transformations. You may want to review some of those topics prior to using this activity.

TRY EXISTING METHODS

Q1 You can't use the sine, cosine, or tangent ratios directly because they apply to right triangles, and this triangle is not a right triangle.

Q2 The Law of Sines is insufficient because one value for each of the three ratios is unknown.

$$\frac{\sin A}{a} = \frac{\sin B}{b} = \frac{\sin C}{c}$$

$$\frac{\sin A}{2} = \frac{\sin B}{6} = \frac{\sin 60°}{c}$$

Q3 You cannot use the Pythagorean theorem because this triangle is not a right triangle.

EXTEND THE PYTHAGOREAN THEOREM

Q4 The value of $a^2 + b^2 - c^2$ is zero at first because this triangle is initially a right triangle, and so the Pythagorean theorem applies to it.

Q5 The lengths of sides a and b stay constant because points A and B are attached to circles which are centered on point C. Side c is the only one that changes.

By keeping a and b constant, students can concentrate on the relationship between $\angle C$ and the quantity $a^2 + b^2 - c^2$. This relationship will reveal the connection with the cosine.

Q6 If $m\angle C < 90°$, the value is positive. If $m\angle C > 90°$, the value is negative.

GRAPH

Q7 The point traces a shape similar to a cosine graph, with a maximum at 0° and a minimum at 180°.

Q8 The graph has the same period as the trace, but is much flatter. The graph's maximum value is 1, but the trace's maximum value is 12.

Q9 To match the trace, you need to stretch the function vertically, increasing its amplitude by a factor of 12.

Q10 Changing d results in a vertical stretch or shrink. To try smaller increments of d, students can double-click the parameter, or they can change its properties to use a keyboard adjustment value of 0.1.

Q11 To match the function to the trace, d must be 12.

Students collect five rows of data. Be sure that they erase traces between trials and that they accurately match the trace and the function plot before recording data.

Q12 Based on the table, $d = 2 \cdot a \cdot b$. The new function is $f(x) = 2ab \cos x$.

Q13 Solving for c^2:

$$2ab \cos C = a^2 + b^2 - c^2$$

$$c^2 = a^2 + b^2 - 2ab \cos C$$

EXPLORE MORE

Q14 Because $\cos(90°) = 0$, it simplifies to $c^2 = a^2 + b^2$.

Q15 To find $\angle C$, given a, b, and c:

$$\cos C = \frac{a^2 + b^2 - c^2}{2ab}$$

You can use this presentation sketch to gather and analyze data from a triangle in order to discover the Law of Cosines.

1. Open **Law of Cosines Present.gsp.** On page 1, begin with three questions.

Q1 Why can't you use the sine, cosine, or tangent functions to find the missing measurements in this triangle? (The triangle is not a right triangle.)

Q2 Write down the Law of Sines and substitute into it the measurements from this triangle. Can you use this equation to solve for the missing side or angles? Explain. (No, each ratio in the Law of Sines still contains an unknown quantity.)

Q3 Can you use the Pythagorean theorem to find the missing side? Give a reason. (No, the Pythagorean theorem works only for right triangles.)

This activity is devoted to finding a new method that does work for triangles like this.

2. Go to page 2 and tell students that they'll begin by calculating a quantity related to the Pythagorean theorem. Use the buttons to measure the sides and to measure $\angle C$.

> We use this name because the expression is the difference between the two sides of the Pythagorean theorem.

3. Calculate $a^2 + b^2 - c^2$, the *Pythagorean difference*.

Q4 Why is this value 0 at first? (The triangle starts with a right angle.)

Q5 Drag A or B to observe how the Pythagorean difference changes as $\angle C$ changes. When is it positive? When is it negative? When is it zero?

4. Plot the Pythagorean difference as a function of $\angle C$. Animate A to fill it in.

Q6 What function does this remind you of? (the cosine function)

5. On page 3, show the plot of the Pythagorean difference. Then show the cosine function.

Q7 The cosine function is too flat. How can you transform it to match the other plot? (with a vertical stretch)

6. Show the transformed function, and adjust d until the curves match.

7. Show the table and double-click it to record the current values. Then change a and b, match again, and record more values. Repeat.

Q8 How is d related to a and b? ($d = 2ab$) Go to page 4, show the table there, and ask if the relationship holds.

8. On page 4, substitute and solve in order to arrive at the Law of Cosines.

9. Finish by asking students to write the two other forms of the Law of Cosines by using $\angle A$ or $\angle B$ in place of $\angle C$.

Probability and Data

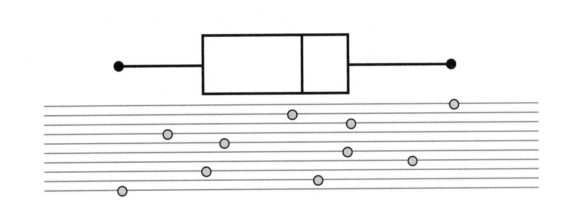

Normal Distribution

Although statistical analysis uses mathematics, it is a science in that it involves the analysis of observations. Recent computer advances have made statistical analysis much easier and more efficient. Statistics is now an indispensable tool in such diverse fields as medicine, biology, economics, sociology, meteorology, and sports.

RANDOM SIMULATION

Suppose you are taking a test, but you are entirely unprepared. In fact, you have no knowledge of the subject matter at all. It's a true/false test with 100 questions, and you need to get at least 60 right to pass. You know that there are 200 students taking that same test, and not one of them knows anything about the subject.

Q1 None of this worries you. Anyone can answer half of the questions correctly simply by guessing, so you only need to be a little bit luckier than most. What would you guess is your probability of getting a passing score? Express your guess as a decimal number between 0 and 1.

1. Open page 1 of **Normal Distribution.gsp.** This sketch will display a random distribution of scores on the test. You can control three parameters:

   p, the probability of a correct answer on any one problem

   pass mark, the minimum score needed to pass the test

   n, the number of people taking the test

2. Edit these three parameters to model the true/false test described above. To try a new random sampling, select the red circle interior and press the exclamation point (!) key. Do this several times, and as you do so, watch the number of passing scores.

> Double-click on a parameter to edit it.

Q2 Based on these observations, about how many people out of 200 are likely to pass the test? What is the approximate probability of passing?

3. Select the red circle interior and press the exclamation point (!) key again several times. This time watch the overall shape of the distribution.

This is something you may have noticed in other data distributions. It has a bell shape, high in the middle and tapering down toward the extremes. Early investigators of probability theory noticed it too. They were successful in deriving a function that would predict the distribution.

4. Open page 2. The curve you see is a prediction of the data distribution. The vertical scale in this model is automatically adjusted so that the curve stays the same when you add more data.

The curve is actually growing taller when you increase *n*. The variable scale makes it appear stable and keeps it from outgrowing the screen.

5. Select the parameter *n* and press the + key several times. This will increase *n* by 100 each time. Run the number up to at least 1000.

Q3 What do you notice about the relationship between the random data and the curve as you increase *n*?

For large values of *n*, this function can give you a reliable approximation of the data distribution before you even see the data.

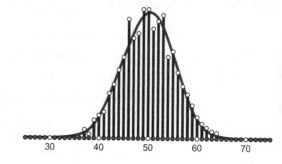

6. Press *Hide Data, Show Total Region,* and *Show Success Region.* When the regions appear, you will also see measurements representing their areas.

Q4 Based on the way you have seen the data fit the graph, how can you use the two area measurements to approximate the probability of passing the test? Do so now. What is the approximate probability?

THE NORMAL DENSITY CURVE

If you have a reasonable level of confidence in the shape of the distribution, you can estimate the probability of an event by comparing areas, just as you did above. The *normal density curve* is actually a family of curves. It has two parameters: μ (mu), the mean; and σ (sigma), the standard deviation. Here is the function definition:

Perhaps you are not yet familiar with *e*. For now, you only need to know that it is a constant. Its value is about 2.72.

$$f(x) = \frac{e^{-((x-\mu)^2/2\sigma^2)}}{\sqrt{2\pi} \cdot \sigma}, \sigma > 0$$

7. Open page 3. This is a normal density curve. Experiment with changing parameters μ and σ.

Q5 Describe in detail how changing μ and σ affects the shape of the curve. Does it cross the *x*-axis?

To set the limits more precisely, edit the parameters *set a* and *set b*, and then press *Set Limits a and b.*

8. Press the *Show Limits* button. The points *a* and *b* on the *x*-axis control the limits of a region under the curve. Notice the measurement for the area of this region. This time it has no units because it is based on a coordinate system.

Exploring Algebra 2 with The Geometer's Sketchpad
© 2007 Key Curriculum Press

For the upper end of this interval, simply drag b to the edge of the screen. The missing area is negligible.

Q6 With the earlier curve the height was based on the size of the data sample. That's not the case here. Drag the region limits to the edge of the screen left and right. What is the approximate total area under the curve? Try this again for several μ and σ settings. How does this simplify the probability calculation?

Q7 The true/false test example has normal parameters $\mu = 50$ and $\sigma = 5$. Enter these parameters and adjust the limits so that the region covers $x > 59.5$. What is your probability estimate?

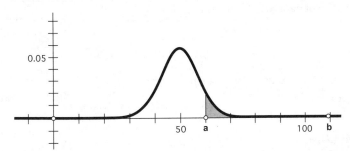

Q8 If the test contained 50 questions worth 2 points each, the parameters would be $\mu = 50$ and $\sigma \approx 7.07$. In that case, would you be more likely or less likely to pass the test?

EXPLORE MORE

It was possible to calculate μ and σ for these true/false test examples, but random variables are usually much more complex. Take a person's weight as an example. It is influenced by age, diet, health, and a combination of genetic material from thousands of generations of ancestors. No one could possibly gather and process so much information. What you have to do instead is to make measurements of the population or, more likely, a sample of the population. You can then derive μ and σ from that.

Sports is an excellent field for investigating statistics because there is plenty of readily available data on athletes' heights, weights, ages, performances, and even salaries. You will probably have to do your own calculations of the mean and standard deviation.

Q9 The published rosters of six National Basketball Association teams indicate a mean height of 79.4 inches, with a standard deviation of 4.03 inches. Based on this information, what is the probability of a randomly chosen NBA player having a height between 70 and 74 inches?

Normal Distribution

Objective: Students simulate a random distribution of test scores and discover properties of the normal density curve.

Student Audience: Algebra 2/Precalculus/Statistics

Prerequisites: An understanding of simple probability and the mean is necessary. The standard deviation is used here, but no formula is given. Students can use the given σ values.

Sketchpad Level: Easy. The sketches are complex, but students only need edit parameters and click action buttons.

Activity Time: 30–40 minutes

Setting: Paired/Individual Activity (use **Normal Distribution.gsp**)

RANDOM SIMULATION

(The test score is what's called a *binomial random variable*, and it results in a *binomial distribution*. Students don't need to be familiar with these terms to do the activity.)

Q1 The probability of scoring 60 or more is approximately 0.028. Don't expect students to calculate this; the question is only a prompt for discussion.

Q2 This question does not ask for a precise calculation either, but students should be able to give a refined answer based on the observations. Encourage them to keep track of results of new random samples.

The expected number of passing scores is about 6, which makes the probability about 0.03. Students may find this counterintuitive. Although scoring 6 out of 10 is no great feat, scoring 60 out of 100 is highly unlikely.

Q3 As *n* increases, the data distribution tends to fit the curve more closely.

Q4 To approximate the probability, calculate the ratio of the areas:

$$\text{probability} \approx \frac{\text{success area}}{\text{total area}} \approx 0.03$$

This method uses a continuous integration although the random variable is discrete. The regions represent not only the integers between 0 and 100 (the only possible scores), but also all real numbers between the integers and even a very small area outside of the range. This may cause some confusion. Remind

students that it is only an approximation. There is some justification for changing the minimum score to 59.5, and that would make a slight difference in the result.

THE NORMAL DENSITY CURVE

Q5 Changing μ causes a horizontal translation of the curve, and μ corresponds to the score at the maximum point of the curve. Increasing σ stretches the curve horizontally and compresses it vertically.

The curve is above the *x*-axis for any real *x*. That can be confirmed from the function definition. Students may be intimidated by the complexity of the definition. Help them break it into manageable parts. Since σ is a positive number, the denominator must be positive. The numerator is a positive base, *e*, with an exponent, so it must be greater than zero no matter what the exponent is.

Q6 No matter what the settings are for parameters μ and σ, the total area under the curve is 1. That corresponds to the denominator of the area ratio from Q4. Therefore, the probability is simply the area of the success region.

Q7 This is the same probability again, 0.03. Students should set limit *a* to 59.5 and drag *b* far to the right. In theory there should be no upper limit, but that is not possible here, and the missing area is negligible.

This may raise some questions about σ. Here is the formula:

$$\sigma = \sqrt{kp(1 - p)}$$

where *p* is the probability of success on any one problem, and *k* is the number of problems on the test.

Q8 You would be more likely to pass. Increasing σ flattens out the curve, forcing more area into the success region. The logical place to set the lower limit is 59 (since only even scores are possible), and the probability of passing is now about 0.10.

EXPLORE MORE

Q9 Based on the data, the probability is about 0.08. Students should set the parameters $\mu = 79.4$, $\sigma = 4.03$, $a = 70$, and $b = 74$.

Permutation and Combination

You probably learned to count before you even started school, but the method you learned (1, 2, 3, . . .) can be limiting. For example, how many seating arrangements are possible for your classroom? Don't answer that now. Just imagine how long it would take to list them all and count. Permutation and combination formulas help us to count things without actually having to see them.

PERMUTATION

Ms. Caba is giving a small prize to each of the top three students in her class. Each prize is different, and she is distributing them randomly among the winners. How many ways are there to distribute three prizes among three students? Each arrangement is a *permutation*.

1. Open **Permutation and Combination.gsp.** The first page, labeled Practice, has a set of icons grouped in a box. Each icon represents a different prize. The buttons allow you to adjust the number of prizes between zero and five.

2. Set the number of prizes to three. Select a prize and drag it out of the box. A copy of it will remain. Drag one of each prize and arrange them in a row.

3. Again, drag one of each prize out of the box and arrange them in another row, but this time in a different order. Continue forming rows of the three prizes until you have formed every possible arrangement.

Q1 How many arrangements were you able to form? In other words, what is the number of permutations for three distinct objects drawn from a set of three?

Q2 Without counting, how many arrangements do you think you could form with four prizes? (*Hint:* How many arrangements are there in which the first prize is a triangle?)

4. Open the Factorial page. Adjust the number of prizes to three, and press the *List Permutations* button. It will answer Q1 and display the possible arrangements, six in all.

Note that the permutations are listed in a logical order. For the first position there are three prizes to choose from. No matter what you choose for the first, there are two prizes available for the second position. That leaves only one remaining for the last position. Hence, the number of permutations is $3 \cdot 2 \cdot 1 = 6$. This number is written 3! (three factorial).

5. This page will list the number of permutations for any set of prizes numbering between zero and five. Try them all.

Q3 As you can see, the number of permutations gets out of hand in a hurry, which is why this demonstration has an upper limit. How many permutations are there for eight distinct objects drawn from a set of eight?

Q4 For any positive integer n you can calculate $n!$ by taking the product of all positive integers less than or equal to n. But what about zero? Why is $0! = 1$?

SUBSETS

In his class Mr. Brownlow has decided to give a prize to anyone who scores 100% on the final exam. He has wrapped four different prizes, but only two students have qualified. In this case you are still counting permutations, but you are not using up all of the available prizes.

Q5 Given a set of four prizes, you must choose two. How many prizes are available for the first student? How many are available for the second? What is the number of permutations for two objects chosen from a set of four?

The number of permutations for r objects chosen from a set of n is written $_nP_r$. You can compute it from the following formula:

You may also see this written as P_r^n or $P(n, r)$.

$$_nP_r = \frac{n!}{(n-r)!}, \quad \text{where } n \text{ and } r \text{ are integers and } n \geq r \geq 0$$

6. Go to the page labeled Permutation. This allows you to change the n and r parameters in a permutation calculation, again with an upper limit of five. List the permutations and check your answer to Q5. Experiment with other settings.

Q6 You may observe the fact that $_nP_0 = 1$ for any non-negative integer n. Explain why this is true.

Q7 It is also true that $_nP_n = {}_nP_{n-1}$. Explain why.

COMBINATION

Mr. Bozich promised that at the end of the semester he would give a prize to every student who scored 100% on any test. Stacey was the only student who accomplished this feat, and she did it three times. Mr. Bozich has five different prizes wrapped, and he tells Stacey to choose three.

In this case the order of the selections does not matter, because they are all going to the same student. Here you are not counting permutations; you are counting *combinations*.

7. Stay on the Permutation page for now. List the permutations for $_5P_3$. There should be 60.

Q8 The first permutation has a circle, a square, and a star. Looking carefully, you can see that six of the permutations have this same combination of prizes. Find them. Explain why there must be six.

You cannot count this same combination six times, and the same goes for all of the other combinations. Therefore, divide the number of permutations by six to get the number of combinations. The expression $_nC_r$ represents the number of possible combinations of r objects chosen from a set of n.

The combination may also be written C_r^n, $\binom{n}{r}$, or $C(n, r)$, and is often pronounced "*n* choose *r*."

$$_nC_r = \frac{_nP_r}{r!} = \frac{n!}{r!(n-r)!}, \quad \text{where } n \text{ and } r \text{ are integers and } n \geq r \geq 0$$

8. Go to the page labeled Combination. List $_5C_3$. Experiment with other combinations.

Q9 What are $_nC_0$, $_nC_1$, and $_nC_n$?

Q10 Explain why $_nC_r$ is always the same as $_nC_{n-r}$.

CALCULATIONS

9. Open the page labeled Calculations. This page has calculations for factorial, permutation, and combination. There are no graphical representations, but there are also no upper limits for the parameters.

Q11 A basketball team has ten players, and there are five different player positions on the floor. How many different starting lineups are possible?

Q12 Nine apartment tenants all drive, and their parking lot has only nine spaces. How many ways are there to arrange the cars in the spaces?

Q13 At the start of the game of euchre, each player gets five cards from a deck that includes only the cards from nine up to ace (9, 10, J, Q, K, A) in the usual four suits. How many hands are possible in euchre?

Permutation and Combination

Objective: Students manipulate a prepared sketch to model permutations and combinations drawn from a given set of objects.

Student Audience: Algebra 2/Statistics

Prerequisites: This activity could serve as an introduction to permutation and combination, but it would work better as a review. At the least, students should understand factorial notation before doing this.

Sketchpad Level: Easy. Students only need to press buttons on the pre-made sketch.

Activity Time: 40–50 minutes

Setting: Paired/Individual Activity (use **Permutation and Combination.gsp**)

PERMUTATION

Q1 There are six ways to arrange the prizes.

Q2 By adding one more prize, there are now 24 ways: For each prize chosen first, there are 6 ways to arrange the remaining three prizes.

Q3 For eight objects the number is $8! = 40,320$.

Q4 We usually say that $0! = 1$ by definition, for convenience, but there is a perfectly logical way of justifying this. Imagine starting with an empty set. If you do nothing at all, then you have one arrangement, and there is no way to change that arrangement. Therefore, there is one way to arrange zero objects.

SUBSETS

Q5 There are four prizes available for the first prize. After choosing that, there are three available for the second. There are 12 permutations.

Q6 Here is a symbolic proof that $_nP_0 = 1$:

$$_nP_0 = \frac{n!}{(n-0)!} = \frac{n!}{n!} = 1$$

Another explanation is that $_nP_0$ is the number of ways to draw zero objects from a set of n. There is one way to do this, which is to draw nothing at all.

Q7 Symbolically:

$$_nP_n = \frac{n!}{(n-n)!} = \frac{n!}{0!} = \frac{n!}{1} = \frac{n!}{1!} = \frac{n!}{[n-(n-1)]!} = {_nP_{n-1}}$$

Imagine drawing n objects from a set of n. You start by drawing $n - 1$ objects. There are $_nP_{n-1}$ ways to do this. In each case there is only one object left to complete the group.

COMBINATION

Q8 There must be six because that is the number of ways there are to arrange three objects ($3! = 6$).

Q9 $_nC_0 = 1$ $_nC_1 = n$ $_nC_n = 1$

Q10 Symbolically:

$$_nC_r = \frac{n!}{r!(n-r)!} = \frac{n!}{[n-(n-r)]!(n-r)!} = {_nC_{n-r}}$$

When you draw r objects, you could think of that as separating the set into two groups, the r objects that you drew and the $n - r$ objects that you left behind. If $_nC_r$ is the number of ways to draw r objects, then it must also be the number of ways to leave $n - r$ objects behind.

CALCULATIONS

Q11 You are drawing five players from a set of ten, and their positions do matter, so use the permutation formula.

$$_{10}P_5 = 30,240$$

Q12 This is simply the number of ways to arrange nine objects.

$$9! = 362,880$$

Q13 Each player gets five cards from a set of 24 different cards, and the order of the cards does not matter.

$$_{24}C_5 = 42,504$$

Box and Whiskers

The *box-and-whiskers plot* (sometimes just called a *box plot*) is a recent development in statistical analysis. You cannot derive any detailed information from it, but it gives you a convenient, easily understood graphical representation of the data distribution.

SKETCH AND INVESTIGATE

1. Open **Box and Whiskers.gsp.**

The sketch contains ten data values represented as points on parallel lines. Above the points are a box and whiskers. You can change a value by sliding its corresponding point right or left. The data are ordered and displayed on the left, but the actual numerical values are not important for this activity.

2. Before answering any of the questions, take a minute to experiment with the sketch. Drag the data points and observe the effect.

Each of the following questions suggests a special shape for the box and whiskers. In each case, state whether it is possible. If it is not possible, explain why not. If it is possible, make a rough sketch of the data points that will create that configuration, and suggest a real data set that might make this happen. The first one is done as an example.

Q1 Can one whisker have zero length?

A1 This will occur if the lower one-fourth of the data points all have the same value. This might happen if there is a lower limit to the data range. Test a group of people to see how far they can throw a heavy weight. Those who cannot even lift the weight will all score zero.

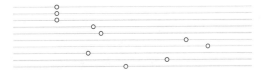

Q2 Can the median fall outside of the box?

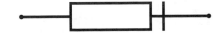

Q3 Can the box have zero width?

Q4 Can both whiskers have zero length?

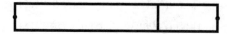

3. Press the *Show Mean* button. The mean is represented by a green bar. Normally, the mean is not shown on a box-and-whisker plot. It appears here so that you can observe its relationship to the data distribution.

Q5 Can the mean fall outside of the box?

Q6 Can the mean be greater than the maximum?

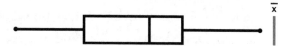

Questions Q7–Q9 involve moving data points. In doing so, you are actually moving from one data set to another. In real life this could happen when a restaurant manager changes the prices of certain menu items, or when a sports team makes a player trade.

Q7 By moving one or more data points, can you move the mean without changing the box, the whiskers, or the median?

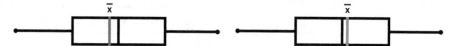

Q8 Can you move a single data point without changing the mean?

Q9 Can you move two data points without changing the mean?

Q10 Can you change the median by moving a single data point?

Exploring Algebra 2 with The Geometer's Sketchpad
© 2007 Key Curriculum Press

Box and Whiskers

Objective: Students manipulate data points and observe the effect of their changes on a box-and-whiskers plot. In the process they gain an intuitive understanding of the box-and-whiskers representation of data distribution.

Student Audience: Algebra 1/Algebra 2/Precalculus/Statistics

Prerequisites: Students should already have seen the box-and-whiskers plot, and should understand the meaning of the five-number summary and the mean of a data sample.

Sketchpad Level: Easy. Students drag points on a prepared sketch; they do no constructions.

Activity Time: 40–50 minutes. The actual computer lab work can be performed in 30 minutes or less, but more time will be needed to answer the open-ended questions.

Setting: Paired/Individual Activity (use **Box and Whiskers.gsp**) or Whole-Class Presentation (use **Box and Whiskers Present.gsp**)

SKETCH AND INVESTIGATE

Q1 See A1 in the student section.

Q2 No, this is not possible. The middle half of the data must fall within the box, and the median must be somewhere in that middle half.

Q3 Yes. It will happen if the middle half of the data have the same value somewhere in the middle of the range.

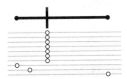

Imagine a math test with only two questions, one very easy and one very difficult. Nearly all of the students would score 1, but a few outliers could score 0 or 2.

Q4 Yes. It is similar to the distribution in Q1, except that there is both a lower limit and an upper limit.

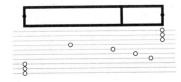

Show a photo of the cast of *Gilligan's Island* to a large group of people. Ask them to identify the characters. Many people have no knowledge of the show, and would score zero. Of the people who have seen the show, most would have no difficulty naming all seven.

Q5 Yes. This can be caused by a fairly close grouping with some extreme outliers.

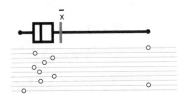

This situation has been modeled in real life in baseball labor disputes. The mean income of professional baseball players is very high. However, most professional players play in the minor leagues and earn a modest income. Relatively few outliers in the majors earn exorbitant salaries, raising the mean but having little effect on the median or the quartiles.

Q6 No, this is not possible. The mean must be within the range of the sample. You can prove this algebraically.

$$x_{max} \geq x_1, x_{max} \geq x_2, \cdots, x_{max} \geq x_{10}$$

$$10x_{max} \geq x_1 + x_2 + \cdots + x_{10}$$

$$x_{max} \geq \frac{x_1 + x_2 + \cdots + x_{10}}{10}$$

$$x_{max} \geq \bar{x}$$

Q7 Yes. When ten data elements are ordered, the five-number summary is defined by only six of the values. The minimum is x_1, the maximum is x_{10}, the first quartile is x_3, the third quartile is x_7, and the median is the mean of x_5 and x_6. Moving any of the remaining four values will change the mean with no effect on the box, provided they do not pass any of these above-mentioned points.

Q8 No. The mean formula includes every data point, so changing a single value always affects the mean.

Q9 Yes. If you move two data points the same distance in opposite directions, the sum stays the same, and so does the mean.

Q10 Yes. The median is the mean of x_5 and x_6, so it is only necessary to move or replace one of these two points.

Box and Whiskers

1. Open **Box and Whiskers Present.gsp.** The sketch has a set of ten data values controlled by ten points on parallel lines. A box-and-whisker summary appears above the points. The actual ordered data are in a column on the left. However, the scale is arbitrary, so you probably will have no use for the actual data.

2. Drag some of the red points to show the class how this affects the image.

Q1 How can you make the box narrower? (Group the data more closely.)

Q2 How can you have a narrow box but long whiskers? (Group most of the data closely, but have one outlier an each side.)

For Q3–Q8, press *Hide Box and Whiskers* before asking. Have students guide your manipulation of the data. Press *Show Box and Whiskers* after the class has reached something approaching a consensus. For each configuration, try to imagine how it could occur with real data. Some suggestions are in the activity notes.

This presentation can be particularly engaging if you give the controls to a student while you and the class direct him or her.

Q3 How can one whisker have zero length? (Group one end of the data on the same value.)

Q4 How can the median fall outside of the box? (This is not possible.)

Q5 Can the box have zero width? (Group the middle part of the data on the same value.)

Q6 Can both whiskers have zero length? (Put the three lowest on the same value and the three highest on the same value.)

3. Press the *Show Mean* button.

Q7 Can the mean fall outside of the box? (Try closely grouped data, but with one extreme outlier.)

Q8 Can the mean be greater than the maximum? (This is not possible.)

For the remaining questions, do not hide the box and whiskers. You will have to be able to see it as the data points are moving.

Q9 Is it possible to move the mean without changing the box, the whiskers, or the median? (Moving any data point will change the mean, but the box and whiskers are unchanged when you move the 2nd, 4th, 7th, or 9th data points.)

Q10 Is it possible to move one point without moving the mean? (It is not.)

Q11 Is it possible to move two points without moving the mean? (Yes, provided they are moved the same distance in opposite directions.)

© 2007 Key Curriculum Press

Fitting Functions to Data

The real world doesn't often behave as cleanly as mathematical functions do. The precision and consistency of pure mathematics is one of its attractions, but also one of its challenges: It's not easy to fit the messy data of real life into the more orderly world of mathematics. In this activity you'll use a general technique for fitting functions (which are smooth and well defined) to data (which may not be).

The technique involves starting with a parent function and using translations, stretches, and shrinks to generate a transformed function that more closely fits the data. To get the best fit, you need to be able to measure how close the fit is. You will use the method of *least squares* for this purpose: You will find the vertical distances from the data points to the function, square those distances, and add them up. The smaller the sum, the better the fit. (This is the reason for the method's name.)

MAKE THE SQUARES

1. Open **Fitting Functions.gsp.**

The sketch has ten data points and four sliders. Your job is to transform a parent function using the sliders so that it fits the data points as closely as possible.

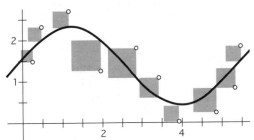

Q1 Looking at the pattern of these data points, what parent function might you use?

2. Press *Show Parent Function* to reveal the parent function to use on this page.

3. Use the parent function and the sliders to graph the transformed function

> To graph g(x), choose **Graph | Plot New Function.** To enter a slider value into the formula, click it in the sketch.

$$g(x) = a \cdot f\left(\frac{x - h}{b}\right) + k$$

Q2 Drag the sliders to change the transformation. Which slider adjusts the horizontal stretch/shrink? Which one adjusts the vertical translation?

4. Arrange the transformed function plot in a rough approximation of the data.

Now you're ready to measure the distances and find the sum of the squares.

> You cannot see the sum yet because it is hidden.

5. Press and hold the **Custom** tools icon and choose **Initialize Function and Sum** from the menu that appears. To start the summation process, click this tool on the transformed function. (Click the function itself not the graph.)

> Press and hold the **Custom** tools icon to see the menu again.

6. To construct the square for the first data point, choose the **Next Square** custom tool from the Custom Tools menu, and click it on the point.

7. Construct squares for each of the remaining data points.

8. To see the sum of the squares, choose the **Show Result** custom tool. You don't even have to click this tool; it shows the sum as soon as you choose it from the menu.

MAKE THE SQUARES SMALL (LEAST SQUARES)

As you adjust, you'll need to switch back and forth among the different sliders to get the best fit.

Q3 Drag the sliders while you watch the *Sum of Squares* calculation, and try to make this sum as small as possible. When you're satisfied, record the slider values and the *Sum of Squares* that you used.

Q4 Page 2 contains data to fit with a linear function. Transform the parent function, $f(x) = x$. Record the slider values and the sum of squares that gives the best fit.

To make very small slider adjustments, select the point on the slider and press the right or left arrow key on the keyboard.

Q5 How many sliders do you really need in order to adjust a linear function? Explain why you don't need all four transformations in this case.

Q6 Page 3 contains data to fit with a quadratic function. The parent function is $f(x) = x^2$. Chris adjusted the sliders to make the sum of the squares 12.10. Can you do better? Record the slider values and the sum of squares that gives the best fit.

Q7 How many sliders do you really need in order to adjust a quadratic function? Explain why you don't need all four transformations in this case.

Q8 Page 4 contains yet more data and an exponential parent function. Fit the transformed function to the data and record your results.

Q9 Why do you think we square the distances from the point to the function before adding them up? Why don't we just find the sum of the distances and use that to measure how good the fit is?

EXPLORE MORE

Q10 Pages 5 and 6 contain additional data, but no functions. On these pages you must create your own parent function, transform it, and fit it to the data points. Choose your function carefully; for some data you may want to try two different parent functions to see which one fits best. (To do a second least-squares calculation in the same sketch, start over with the **Initialize Function and Sum** tool, and then use the **Next Square** tool on each data point.)

9. Collect some data of your own. The data can come from your own measurements, from a science lab, from the Internet, or from some other source. Plot the data in Sketchpad, and then choose an appropriate function and fit it to the data. Present the results to your group or to your entire class.

Fitting Functions to Data

Objective: Students transform functions to fit data, using a least-squares calculation to judge how good the fit is. They use several different parent functions, depending on the shape of the data.

Student Audience: Algebra 2/Precalculus/Statistics

Prerequisites: Students must be familiar with function transformations (translations and stretches/shrinks). It's useful if they also have some experience with fitting linear functions to data.

Sketchpad Level: Intermediate. There's not much construction, but students must know how to create and graph functions. Students use custom tools to perform the least-squares calculations.

Activity Time: 35–45 minutes

Setting: Paired/Individual Activity (use **Fitting Functions.gsp**) or Whole-Class Presentation (use **Fitting Functions Present.gsp**)

MAKE THE SQUARES

Q1 The data appear to be periodic, so the $\sin x$ or $\cos x$ function may be a good choice for the parent function.

Q2 Slider b adjusts the horizontal stretch/shrink. Slider k adjusts the vertical translation.

5. If students are not familiar with the Custom Tools menu, show them how to press and hold the **Custom** tools icon to make it appear.

MAKE THE SQUARES SMALL (LEAST SQUARES)

Answers will vary for the questions asking students to actually fit functions. Some typical values for Q3, Q4, Q6, Q8, and Q10 are shown in the table below, but there is the possibility for considerable variation. For instance, the sine function used in Q3 will require very different values for the horizontal translation (h) if the vertical stretch (a) is negative instead of positive.

Q5 You need only two of the sliders for a linear function. The values of a and b combine to determine the slope. The values of h and k combine to determine the intercept.

Q7 You need only three of the sliders for a quadratic function. The values of a and b combine to determine the width of the parabola. (The values of h and k determine the vertex.) Similarly, square root and absolute value functions need only three sliders.

Q9 One reason for squaring the values is to make them all positive. If you just add the deviations, a large negative deviation and a large positive one might add up to zero, even though both points are far from the graph.

Q3 (page 1)	$y = 2.05 \sin\left(\frac{x - 5.20}{1.45}\right) + 0.90$
Q4 (page 2)	$y = -0.60x + 2.40$
Q6 (page 3)	$y = -\left(\frac{x - 14.95}{3.20}\right)^2 + 25.25$
Q8 (page 4)	$y = 50.25 \cdot 2^{-(x-0.25)/12.60} - 0.25$
Q10 (page 5)	$y = 1.25\sqrt{x - 7.90} + 0.15$
Q10 (page 6)	$y = \frac{7.5}{x - 5.00} + 1.05$

WHOLE-CLASS PRESENTATION

Use the Presenter Notes and **Fitting Functions Present.gsp** to present this activity to the whole class.

Use this presentation to review function transformations and to introduce students to some of the principles of curve fitting.

MAKE THE SQUARES

1. Open **Fitting Functions Present.gsp.** The sketch has ten data points and four sliders. Tell students that their job is to transform a parent function using the sliders so that it fits the data points as closely as possible.

Q1 Looking at the pattern of these data points, what sort of parent function might you use? (The data look periodic, so $\sin x$ or $\cos x$ might be a good choice.)

2. Press *Show Parent Function* to reveal the parent function for this page.

3. Press *Show Transformed Function* to create and graph the transformed function.

$$g(x) = a \cdot f\left(\frac{x-h}{b}\right) + k$$

It's best to have a student operating the computer and taking direction from you and from other students.

Q2 Ask students how to drag the sliders to experiment with the transformation. You're not trying to fit the data yet; you are familiarizing students with the effects of the sliders. Students should drive this adjustment process. Don't rush them; give them time to discuss, to argue, and to think about what's happening.

Q3 Which slider adjusts the horizontal stretch/shrink? (*b*)

Q4 Which slider adjusts the vertical translation? (*k*)

4. Adjust the sliders to arrange the transformed function plot in a rough approximation of the data.

Now you're ready to measure the distances and find the sum of the squares.

5. Use the **Initialize Function and Sum, Next Square,** and **Show Result** custom tools to find the sum of the squares of the deviations. Consult the directions for the student activity for details on using these tools.

To make very small slider adjustments, select the point on the slider and press the right or left arrow key on the keyboard.

Q5 Adjust the sliders to minimize the sum of the squares. Switch back and forth among the various sliders to be sure you've found a minimum result. Record the slider values.

Q6 Why do you think we square the distances from the point to the function before adding them up? Why don't we just find the sum of the distances and use that to measure how good the fit is? (We use squared values to make all the values positive so that positive and negative deviations do not cancel each other out.)

The remaining pages contain a number of interesting data patterns to fit.

Vectors and Matrices

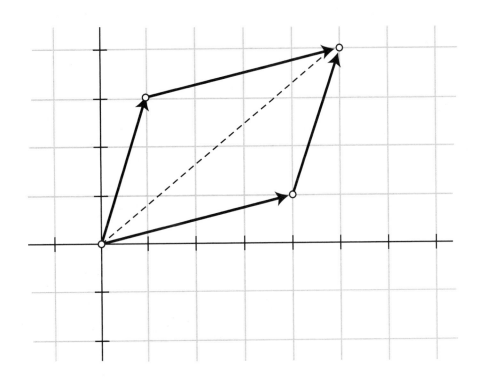

Introduction to Vectors: Walking Rex

You know that $2 + 2 = 4$. But if you walk 2 miles north and then 2 miles south, how far did you go? In one sense you went 4 miles—that's what your feet would tell you. But in another sense you haven't gone anywhere. We could say: $2N + 2S = 0$.

Values with both magnitude (size or length) and direction are called *vectors*. Vectors are useful in studying the flight paths of airplanes in wind currents and the push and pull of gravitational or electric forces. In this activity you'll explore the algebra and geometry of vectors in the context of a walk with your faithful dog, Rex.

WALK THE DOG

1. Open **Introduction to Vectors.gsp.** Rex's leash is tied to a tree at the origin of an *xy*-coordinate system. Rex is pulling the leash tight as he excitedly waits for you to take him on a walk.

Vectors often have a physical meaning. This particular vector represents Rex's position relative to the tree.

Rex's taut leash is represented by a vector, a segment with an arrowhead. The end with the arrow (Rex) is called the *head*, and the plain end is called the *tail*. We've labeled this particular vector **a**.

Q1 One way to define a vector is by its magnitude and direction. Which of these two quantities stays the same as you drag point *Rex*?

Another way to define a vector is by the coordinates of its head when its tail is at the origin. These coordinates are called the *components* of the vector.

You will not be able to make the components match the given values exactly. Just make them as close to those values as you can.

Q2 For each problem, drag *Rex* so the vector has the given components, and find the magnitude and direction of the vector.

    a. components are $(5, 0)$

    b. components are $(-4, 3)$

    c. components are $(0, -5)$

    d. components are $(-3, -4)$

Q3 For each problem, drag the vector as close as you can to the given magnitude and direction, and find the vector's components.

  a. magnitude = 5; direction = 30°

  b. magnitude = 5; direction = 135°

  c. magnitude = 5; direction = 240°

  d. magnitude = 5; direction = 307°

Q4 Rex is terrified of ladybugs. Suppose a ladybug is sitting at $(5, 0)$. Where should Rex move to face in the opposite direction and be as far from it as possible? Describe Rex's position both ways, using components and using magnitude and direction.

Q5 What if the ladybug moves to a position 5 units away from the tree at 140°? Where should Rex go now?

Now it's time to untie the leash from the tree and take Rex for a walk.

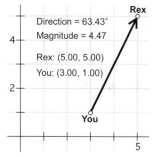

2. Go to page 2. Rex is very determined! As you walk him, he pulls the leash taut and always tries to pull you in the same direction. Rex is still at the head of vector **a**, and now you're at the tail.

Q6 What are the components of vector **a**? How can you determine the components without moving point *You* to the origin?

Q7 Drag vector **a** around the screen. Explain why, no matter where you drag it, vector **a** is always the same vector. Support your argument using both of the two methods for describing vectors.

Q8 Suppose you walk to the point $(80, 80)$. Where will Rex be? Explain how you found your answer. (Don't scroll—all the information you need is on the screen.)

3. Go to pages 3 and 4. Rex is heading in different directions on these pages. The information presented on screen is also a little different for each page.

Q9 On each of these two pages, determine where Rex will be standing when you're at $(80, 80)$. Explain your reasoning in each case.

Q10 What if your leash is twice as long, and Rex is still pulling in the same direction? Now where will Rex be when you're at $(80, 80)$? Answer for both page 3 and page 4.

Exploring Algebra 2 with The Geometer's Sketchpad
© 2007 Key Curriculum Press

Introduction to Vectors: Walking Rex

Objective: Students learn two ways to describe vectors, convert between the two descriptions, and move vectors around to explore how the vector is independent of any specific position.

Student Audience: Algebra 1/Algebra 2

Prerequisites: None. This activity is designed to be a first introduction to vectors.

Sketchpad Level: Easy. Students work with a pre-made sketch.

Activity Time: 20–30 minutes. It may be possible to do this activity and the follow-up activity, Vector Addition and Subtraction, in one class period.

Setting: Paired/Individual Activity (use **Introduction to Vectors.gsp**) or Whole-Class Presentation (use **Introduction to Vectors Present.gsp**)

Related Activity: Vector Addition and Subtraction

WALK THE DOG

Q1 The magnitude stays the same. (In other words Rex is always the same distance—the length of his leash—from the tree.)

Q2 Students should be able to get within about 1° of the answers below.

    a. magnitude = 5; direction = 0°

    b. magnitude = 5; direction = 143.13°

    c. magnitude = 5; direction = 270°

    d. magnitude = 5; direction = 233.13°

Q3 Students should be able to get within about 0.1 of the answers below.

    a. (4.33, 2.50)

    b. (−3.54, 3.54)

    c. (−2.50, −4.33)

    d. (3.00, −4.00)

Q4 To get away from the ladybug when it's at (5, 0), Rex should move to (−5, 0). In magnitude-direction form he should move 5 units in direction 180°. In general, vectors (x, y) and $(−x, −y)$ face in opposite directions.

Q5 To get away from the ladybug when it's at distance 5 units and direction 140°, he should move to distance 5 units and direction 320°. In component form he should move to (3.83, −3.21). Assuming positive magnitude, a vector in direction θ and a vector in direction $\theta + 180°$ face in opposite directions.

Q6 The components of **a** are (2, 4). To find these components without moving point *You* to the origin, you can subtract the coordinates of the tail from the coordinates of the head.

Q7 The only things that change when vector **a** is dragged are the locations of its head and tail. The first method for describing vectors uses magnitude and direction—neither of these changes as **a** is dragged. The second uses the components, and these don't change either. No matter where the head and tail actually are, the result of the subtraction never changes.

Q8 (82, 84). No matter where you're standing, Rex is 2 units to the right and 4 units up, because the components of the vector are (2, 4).

Q9 Page 3: (83, 81). Find the components by dragging the vector so its tail is at the origin. The components are (3, 1), so Rex is always 3 units to the right of you and 1 unit up from you.

Page 4: (74, 83). Find the components by dragging the vector so its tail is at the point marked (8, 6) and then subtracting. The components are (−6, 3), so if you are at (80, 80), Rex will be at (74, 83).

Q10 Page 3: (86, 82). If the leash is twice as long, the components must be (6, 2).

Page 4: (68, 86). If the leash is twice as long, the components must be (−12, 6).

Introduction to Vectors: Walking Rex

Use this presentation to introduce vectors and two common ways to describe them. Vectors often represent physical quantities, and the scenario of walking a dog on a leash provides a hook students can use to understand the concepts better.

Q1 If you walk 2 miles north and then 2 miles south, how far did you go? (Encourage discussion of the alternative answers: 4 miles or no distance at all. Use the discussion to point out that distance alone isn't enough here, and that you also have to take into account the direction of the walk.)

Define a *vector* as a quantity that has both magnitude (size or length) and direction, and describe some areas in which vectors are important in everyday life.

WALK THE DOG

This activity is particularly effective with a student operating the computer and taking direction from you and from other students.

1. Open **Introduction to Vectors Present.gsp.** Describe the situation: Rex's leash is tied to a tree at the origin of an *xy*-coordinate system. Rex is pulling the leash tight as he excitedly waits for you to take him on a walk.

Q2 Get students to answer each of the questions on this page. Remind them periodically of the equivalence of the two different ways they are using to describe the same object: by components or by magnitude and direction.

Now it's time to untie the leash from the tree and take Rex for a walk.

2. On page 2 you've untied Rex from the tree, and he's pulling in a certain direction, no matter where you try to drag your end of the leash.

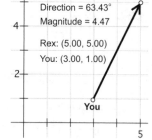

Q3 What are the components of vector **a**? How can you determine the components without moving point *You* to the origin?

Q4 Drag vector **a** around the screen and ask whether the vector changes as the coordinates of its head and tail change. Get students to explain why, no matter where you drag it, vector **a** is always the same vector. Encourage them to use both methods of describing the vector in their arguments.

Q5 On page 3, where will Rex be when you're at (80, 80)? How can you figure this out? Let the discussion develop until students find and understand a strategy.

Q6 On page 4 the objective is the same, but the available information is different. Encourage students to describe their strategy to each other.

Q7 What if your leash is twice as long, and Rex is still pulling in the same direction? Now where will Rex be when you're at (80, 80)? Answer for page 3 and page 4.

Vector Addition and Subtraction

You may have done an earlier activity in which you learned about vectors as you walked your dog Rex. In this activity you and Rex will go on some more walks together. Along the way you'll explore how to add and subtract vectors.

SKETCH AND INVESTIGATE

1. Open **Vector Add Subtract.gsp.**

Page 1 has two movable vectors, **a** and **b.** You and Rex begin your walk at home (the origin), heading first along a path the direction and magnitude of vector **a,** and then along another path the direction and magnitude of vector **b.**

Magnitude is another word for the *size* or *length* of a vector.

Q1 Drag the vectors to represent your walk. Find the coordinates of the ending point of your walk. Explain how you got your answer.

Q2 Some days Rex prefers a different route: **b,** then **a.** Where do you and Rex end up after taking this route?

Your walks in Q1 and Q2 could have taken more direct paths. Instead of following vectors **a** and **b,** you could have gone directly to the final destination along a single vector, **c.**

Q3 Choose **Vector** from the Custom Tools menu. This tool draws a vector with an arrow at its head. Use it to draw the vector **c** that would take you directly from home to your final destination. What are the coordinates of the head and tail of **c?**

Q4 On page 2, the coordinates of only one point are given. Assuming that you start at home, where do you end up? How did you get your answer?

2. Use the **Vector** tool to draw the vector representing the direct path from home to your final destination.

As before, you won't need to measure any new coordinates.

Q5 On page 3 are two identical pairs of vectors. Build one route to your destination with one set (**a,** then **b**), and then the other route with the other set (**b,** then **a**). For both routes, at what coordinates do you end up? Do the two routes take you to different final destinations? Is one route longer than the other?

3. Use the **Vector** tool to draw the direct path from home to your final destination.

Q6 All along, you've been combining vectors by putting the tail of the second vector on the head of the first. Would you get the correct destination if you put the head of the second vector on the head of the first? Why or why not?

VECTOR ADDITION

What you've just been doing is adding vectors! In step 2 you found a single vector, **c**, that got you to the same place as walking along **a**, then **b**. So we can say that **c** = **a** + **b**.

The coordinate system made it easy to find the coordinates of **c**'s head. But if there is no coordinate grid, can you still draw **a** + **b**?

4. Go to page 4. What do you think **a** + **b** will look like? Make a conjecture, and then press the button *Add a + b* to see.

5. What will **b** + **a** look like? Make a conjecture, and then press *Add b + a* to see.

Q7 On your paper, describe vector addition in a way that someone familiar with the basics of vectors—but not yet vector addition—would understand. Make sure to use the words *head* and *tail*.

Q8 Addition of real numbers is *commutative* because $x + y = y + x$. (For example, $3 + 5 = 5 + 3$. Both operations give the answer 8.) Is vector addition commutative? In other words, does **a** + **b** = **b** + **a**? Justify your answer with a drawing.

Q9 The *zero vector*, **0**, is a vector with a magnitude of zero (basically, just a point). What must be true of two vectors whose sum is **0**?

VECTOR SUBTRACTION

Consider subtraction of real numbers. Recall that by definition, $8 - 5 = 8 + (-5) = 3$. *Subtracting a number is the same as adding its opposite.* The same is true with vectors: *Subtracting a vector is the same as adding its opposite.*

6. On page 5, experiment just as you did on the previous page.

Q10 On your paper, describe vector subtraction in a way that someone familiar with the basics of vectors—but not yet vector subtraction—would understand.

Q11 Is vector subtraction commutative? In other words, does **a** − **b** = **b** − **a**? Defend your answer with a drawing.

EXPLORE MORE

Q12 One way of describing a vector is to use its components: the coordinates of its head if its tail is at the origin. Assume that **a** = (a_1, a_2) and **b** = (b_1, b_2). If **a** + **b** = **c**, what are the components of **c**? If **a** − **b** = **d**, what are the components of **d**?

Exploring Algebra 2 with The Geometer's Sketchpad
© 2007 Key Curriculum Press

Vector Addition and Subtraction

Objective: Students learn to add and subtract vectors graphically and algebraically, and investigate whether these two operations are commutative.

Student Audience: Algebra 1/Algebra 2

Prerequisites: Students should be familiar with the concept of vectors, and should be comfortable describing them using components or using magnitude and direction.

Sketchpad Level: Easy. Students work with a pre-made sketch.

Activity Time: 20–30 minutes. It may be possible to do the activity Introduction to Vectors: Walking Rex and this activity in one class period.

Setting: Paired/Individual Activity (use **Vector Add Subtract.gsp**) or Whole-Class Presentation (use **Vector Add Subtract Present.gsp**)

Related Activity: Introduction to Vectors: Walking Rex

SKETCH AND INVESTIGATE

Q1 The coordinates of the ending point are (7, 9). To find this, put **a**'s tail at the origin and **b**'s tail at **a**'s head. The coordinates of point *B*, **b**'s head, are the answer.

Q2 The answer is still (7, 9).

Q3 The head is at (7, 9) and the tail is at (0, 0).

Q4 You end up at (21, 10). To get there, put **b**'s tail at the origin and **a**'s tail at **b**'s head so that point *A*, whose coordinates are given, is the destination point.

Q5 Both routes take you to (19, 9). Put **a**'s tail at the origin this time and **b**'s tail at **a**'s head so that point *B*, whose coordinates are given, is the destination point. For the other route, put **b**'s tail at the origin and **a**'s tail at **b**'s head. The coordinates for both destinations are the same, so they take you and Rex to the same place. The length of each route is the sum of its component vectors. Since the two routes have the same components, they also have the same length.

Q6 You get an incorrect destination when you line up the head of one vector with the head of another because this violates the meaning of a vector. Vectors have a definite direction. If you line up their heads, you would be going against the direction of the second vector, which is sort of like going the wrong way down a one-way street.

VECTOR ADDITION

Q7 Line up the tail of the second vector with the head of the first vector. The sum is the vector from the first vector's tail to the second vector's head.

Q8 Yes, vector addition is commutative. The drawing on the first page of the activity demonstrates this: Whether you add the second vector to the first or the first to the second, you end up in the same place.

Q9 Two vectors whose sum is zero have the same magnitude but face in opposite directions.

VECTOR SUBTRACTION

Q10 Vector subtraction is just like vector addition except that what is added is the *opposite* of the second vector (which we'll call the *opposite vector* here). Thus, line up the head of the first vector with the tail of the opposite vector. The difference between the two vectors is represented by the vector connecting the tail of the first vector to the head of the second vector's opposite. You are adding the opposite of the second vector to the first vector.

Q11 No, vector subtraction isn't commutative. In fact, the difference of two vectors is the opposite of the difference taken in the reverse order.

EXPLORE MORE

Q12 The components of vector **c** are $(a_1 + b_1, a_2 + b_2)$. The components of vector **d** are $(a_1 - b_1, a_2 - b_2)$.

Vector Addition and Subtraction

Presenter Notes

Substitute the name of one of your students.

In this activity a student, Yolanda, and her dog, Rex, will take some walks together, and in the process explore how to add and subtract vectors.

VECTOR ADDITION

This activity is most effective if a student operates the computer and takes direction from you and from other students.

1. Open **Vector Add Subtract Present.gsp.** Page 1 has two movable vectors, **a** and **b**. Yolanda and Rex begin their walk at home (the origin), heading first along a path the direction and magnitude of vector **a**, and then along another path the direction and magnitude of vector **b**.

Q1 Drag the vectors so they show this walk. Find the coordinates of the ending point of the walk. Have students explain how they got their answers.

Q2 Some days Rex prefers a different route: **b**, then **a**. Rearrange the vectors to represent this route. Where do they end up?

Q3 On page 2 there are two vectors, but only one shows its coordinates. If Yolanda and Rex start at home, where do they end up? How did you have to arrange the vectors?

Choose the **Vector** custom tool by pressing and holding the **Custom** tools icon, and selecting **Vector** from the menu that appears. Use it by clicking first the tail, then the head of the desired vector.

2. Use the **Vector** custom tool to draw the vector representing the direct path from Yolanda's home to the final destination.

Q4 On page 3 are two identical pairs of vectors. Build one route to the destination with one set (**a**, then **b**), and then the other route with the other set (**b**, then **a**). Where does each route end up? Is one route longer than the other?

3. Use the **Vector** tool to draw the direct path from Yolanda's home to her destination.

The new vector you just constructed is the sum of the other two vectors.

Q5 Addition of real numbers is *commutative* because $x + y = y + x$. (For example, $3 + 5 = 5 + 3 = 8$.) Is vector addition commutative? Does $\mathbf{a} + \mathbf{b} = \mathbf{b} + \mathbf{a}$? Justify your answer based on the vector operations you've done.

VECTOR SUBTRACTION

Recall that subtracting a number is the same as adding its opposite. The same is true with vectors: *Subtracting a vector is the same as adding its opposite.*

Q6 How can you make the opposite of a vector? (Keep the same magnitude, but reverse the direction.)

4. On page 4, experiment with vector subtraction. Use the button to find the opposite of **b**. Then subtract the vectors by adding **a** and the opposite of **b**.

Q7 Is vector subtraction commutative? Use the sketch to investigate.

Solving Systems Using Matrices

One of the most common applications of matrices is to solve a system of linear equations. For a solution to be possible, the number of equations must be greater than or equal to the number of unknown values. In this activity you'll solve systems of two equations with two unknowns.

INTERSECTION OF LINES

1. Open **Matrix Solution.gsp**. The sketch contains two lines in the coordinate plane, along with their equations in standard form ($ax + by + c = 0$). You can control the equations and the lines by editing the parameters a_1, b_1, c_1, a_2, b_2, and c_2.

You have seen problems like this before. This is a system of two equations with two unknowns, x and y. The coordinates of the intersection point are the solution. This time you will use matrices to solve the system. First you have to express the system of equations as a matrix equation:

$$a_1x + b_1y = c_1 \atop a_2x + b_2y = c_2 \Rightarrow \begin{bmatrix} a_1 & b_1 \\ a_2 & b_2 \end{bmatrix} \begin{bmatrix} x \\ y \end{bmatrix} = \begin{bmatrix} c_1 \\ c_2 \end{bmatrix}$$

Q1 In this equation there are only two unknowns. What are they?

The next step is to make a formula by solving for the unknowns:

$$\begin{bmatrix} a_1 & b_1 \\ a_2 & b_2 \end{bmatrix} \begin{bmatrix} x \\ y \end{bmatrix} = \begin{bmatrix} c_1 \\ c_2 \end{bmatrix} \Rightarrow \begin{bmatrix} a_1 & b_1 \\ a_2 & b_2 \end{bmatrix}^{-1} \begin{bmatrix} c_1 \\ c_2 \end{bmatrix} = \begin{bmatrix} x \\ y \end{bmatrix}$$

2. This formula requires an inverse matrix. Arrange parameters a_1, b_1, a_2, and b_2 into a matrix as shown here. Press and hold the **Custom** tools icon, and choose **2D Inverse**. Click the four matrix elements, working down the left column first, and then down the right column: a_1, a_2, b_1, b_2.

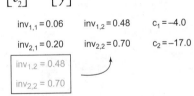

$inv_{1,1} = 0.06 \quad inv_{1,2} = 0.48 \quad c_1 = -4.0$
$inv_{2,1} = 0.20 \quad inv_{2,2} = 0.70 \quad c_2 = -17.0$
$inv_{1,2} = 0.48$
$inv_{2,2} = 0.70$

3. The result of using the tool is the inverse matrix. It appears as a single column. Drag the two bottom elements up and to the right to form the second column. Arrange parameters c_1 and c_2 as a column vector on the right side of the inverse vector.

4. Press and hold the **Custom** tool button, and choose **2D Matrix * Vector**. Select the elements of the matrix formula. Work down the left matrix column first, then down the right column, and finally down the column vector. The output is the solution. The results of this calculation will be labeled *c[1]'* and *c[2]'*. Use the **Text** tool to change the labels to x and y.

5. Select the coordinates in order. Choose **Graph | Plot As (x, y)**.

Q2 Does the plotted point fall on the intersection of the lines?

Q3 By changing the equation parameters, find the solution to this system:

$$3x + 10y = -8$$
$$-4x + y = -18$$

Q4 Edit the parameters again to form the system shown below. The system has no solution. How can you tell that by examining the matrix equation?

$$-12x + 9y = -4$$
$$4x - 3y = -17$$

FITTING A LINE

6. On page 2 there are two points and their coordinates.

As you know, there is exactly one line through both of these points. Your objective is to find the slope-intercept equation of that line. Both points satisfy the equation, so that gives you this system:

$$mx_P + b = y_P$$
$$mx_Q + b = y_Q$$

See those two 1's at the edge of the screen? You will need them when you form your matrix.

Q5 This time you know x and y in both equations. The unknown values are m and b. Rewrite this system of equations as a matrix equation. Then rewrite that as a matrix formula for the unknowns.

7. Repeat the steps from the earlier section to find m and b.

8. Choose **Graph | Plot New Function**. For the function definition, enter $mx + b$.

Q6 If the two points have the same x-coordinate, it is still possible to draw the line between them, but it is not possible to write the equation in slope-intercept form. Drag one of the points so that it is directly above the other. What happens to the function plot? Explain why the matrix solution fails.

EXPLORE MORE

Page 3 is identical to page 2. Use a similar procedure to fit the two points with the graph of an equation in this form:

$$a\cos x + b\sin x = y$$

Solving Systems Using Matrices

Objective: Students model a system of equations as a single matrix equation. They then use the inverse matrix to find the solution.

Student Audience: Algebra 2/Precalculus

Prerequisites: Students should have a basic understanding of matrix algebra. They will not have to calculate an inverse matrix themselves, but they will have to use one.

Sketchpad Level: Intermediate. Students will do most of the work with the aid of custom tools.

Activity Time: 20–30 minutes

Setting: Paired/Individual Activity (use **Matrix Solution.gsp**) or Whole-Class Presentation (use **Matrix Solution Present.gsp**)

Although students should learn how to compute the inverse of a matrix, that task would make this activity more tedious, and it would greatly increase the likelihood of a blunder. For those reasons the partially completed document has custom tools for computing the matrix and for multiplying a matrix by a vector.

INTERSECTION OF LINES

Q1 The unknowns are x and y.

Q2 The plotted point falls on the intersection.

Q3 The solution to the given system is $x = 4$, $y = -2$.

Q4 This system is represented by the following matrix equation:

$$\begin{bmatrix} -12 & 9 \\ 4 & -3 \end{bmatrix} \begin{bmatrix} x \\ y \end{bmatrix} = \begin{bmatrix} -4 \\ -17 \end{bmatrix}$$

The matrix is singular. Students may see that the top row is a multiple of the bottom, or they may see that the determinant is zero. This means that it has no inverse, and the system has no solution. Students may also observe that the lines are parallel.

FITTING A LINE

Q5 Be sure to check for understanding at this point. There may be a tendency for students to try to force the matrix equation into the same form that was used in the previous section, but this situation is very different.

$$\begin{aligned} mx_P + b &= y_P \\ mx_Q + b &= y_Q \end{aligned} \Rightarrow \begin{bmatrix} x_P & 1 \\ x_Q & 1 \end{bmatrix} \begin{bmatrix} m \\ b \end{bmatrix} = \begin{bmatrix} y_P \\ y_Q \end{bmatrix}$$

$$\Rightarrow \begin{bmatrix} x_P & 1 \\ x_Q & 1 \end{bmatrix}^{-1} \begin{bmatrix} y_P \\ y_Q \end{bmatrix} = \begin{bmatrix} m \\ b \end{bmatrix}$$

Q6 This is the same problem that occurred in Q3. The matrix is singular because the top and bottom rows are identical. Therefore, it has no inverse, and the system has no solution. In this case no solution means only that there is no slope-intercept equation. The line may be possible, but its equation would have to be in a different form.

EXPLORE MORE

The trick to this extension is to mentally separate the known values and the unknown. Since the x- and y-coordinates of two points are known, you can also regard the sine and cosine parts of the equation as known values. Use the Sketchpad Calculator to compute $\cos x_P$, $\sin x_P$, $\cos x_Q$, and $\sin x_Q$. Then arrange those calculations into a matrix as shown:

$$\begin{aligned} a\cos x_P + b\sin x_P &= y_P \\ a\cos x_Q + b\sin x_Q &= y_Q \end{aligned} \Rightarrow \begin{bmatrix} \cos x_P & \sin x_P \\ \cos x_Q & \sin x_Q \end{bmatrix} \begin{bmatrix} a \\ b \end{bmatrix} = \begin{bmatrix} y_P \\ y_Q \end{bmatrix}$$

$$\Rightarrow \begin{bmatrix} \cos x_P & \sin x_P \\ \cos x_Q & \sin x_Q \end{bmatrix}^{-1} \begin{bmatrix} y_P \\ y_Q \end{bmatrix} = \begin{bmatrix} a \\ b \end{bmatrix}$$

After solving for a and b, define and plot the function $f(x) = a\cos x + b\sin x$. The result creates a much more interesting graph than the linear function although the procedure is nearly the same.

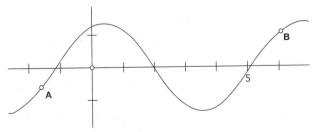

Solving Systems Using Matrices

In this presentation you'll show students how to solve a system of equations by forming a matrix of coefficients and using the inverse matrix to compute the solution.

The presentation will be most effective if you have a student operate the computer.

1. Open **Matrix Solution Present.gsp.** The six red and blue parameters control the two linear equations above them. They also control the lines themselves. Change some of the parameters to show the relationship.

Q1 The solution to this system of linear equations corresponds to the intersection of the lines. How can you represent this system as a single matrix equation? After suitable input from students, press *Matrix* to show the answer.

Q2 The variables x and y are the only unknowns here. How can you rewrite the equation to solve for x and y? Press the *Inverse* button to show the formula.

At this time you may wish to have the class derive the inverse of the matrix. If so, simplify the task by making all the parameters integers.

2. Press *Compute* to show the inverse matrix, and press *Solution* to show the plotted point.

3. Before moving to the next part, press *Reset* and briefly go through the steps again.

4. Go to the Fitting a Line page. This time there are two points and no line.

5. Press *Coordinates*. Drag the points around and explain that the objective is to find the slope-intercept equation of the line joining these two points.

Q3 What two equations can you write using the given information? Tell students to take their time on this part. They may try to force the equations into the same form as in the previous section. Get them to use the slope-intercept form. After a suitable period of discussion, press *Show Equations*.

6. Continue through the buttons in order, showing the matrix, its inverse, and finally the solution. Urge students to keep in mind which values they are looking for. Rather than x and y, they are now looking for m and b.

Q4 After showing the solution, drag one of the points so that it is directly above the other. Why does the solution line disappear? Students may know that the slope-intercept equation cannot represent a vertical line, but how can they tell from the matrix? Use the Calculator to show that the determinant is zero.

Q5 Go back to the Intersection of Lines page. Adjust the parameters to form the equations below. Why is there no solution? Again, show that the determinant is zero.

$$-12x + 9y = -4$$

$$4x - 3y = -17$$

Key Curriculum Press
Innovators in Mathematics Education

Comment Form

Please take a moment to provide us with feedback about this book. We are eager to read any comments or suggestions you may have. Once you've filled out this form, simply fold it along the dotted lines and drop it in the mail. We'll pay the postage. Thank you!

Your Name _____

School _____

School Address _____

City/State/Zip _____

Phone _____

Book Title _____

Please list any comments you have about this book.

_____
_____
_____
_____
_____
_____
_____
_____
_____
_____

Do you have any suggestions for improving the student or teacher material?

_____
_____
_____
_____
_____
_____
_____

To request a catalog, or place an order, call us toll free at 800-995-MATH, or send a fax to 800-541-2242. For more information, visit Key's website at www.keypress.com.

Please detach page, fold on lines and tape edge.

NO POSTAGE
NECESSARY
IF MAILED
IN THE
UNITED STATES

BUSINESS REPLY MAIL
FIRST CLASS   PERMIT NO. 338   OAKLAND, CA

POSTAGE WILL BE PAID BY ADDRESSEE

KEY CURRICULUM PRESS
1150 65TH STREET
EMERYVILLE CA 94608-9740
ATTN: EDITORIAL